j

)6

The Perception of
Light and Colour

The Perception of Light and Colour

C. A. PADGHAM, A.R.C.S., M.SC., PH.D., D.I.C., F.INST.P.
Reader in Physiological Optics

and

J. E. SAUNDERS, M.SC., PH.D., M.INST.P.
Lecturer in Physiological Optics
The Department of Ophthalmic Optics and Visual Science,
The City University, London

LONDON
G. BELL & SONS LTD
1975

ISBN 0 7135 1874 X

Filmset by Typesetting Services Ltd, Glasgow, Scotland

Printed in Great Britain by
Fletcher and Son, Ltd., Norwich

Foreword

When Dr Padgham invited me to write a Foreword to the book on which he was working with Dr Saunders, I was very pleased to accept, for two reasons. In the first place, I have been associated with Dr Padgham for many years, both at Imperial College and the City University, through our common interest in optics and vision. Dr Padgham has made very significant contributions to visual optics through his own researches and also through the encouragement and guidance he has given to his research students. Indeed, his co-author, Dr Saunders, was one of his students and this only increases the pleasure of commending this book to a wide readership.

My second reason for welcoming the opportunity to write the Foreword was the sheer fascination of the subject of the book. The processes of visual perception are so remarkable in their complexity and so effortless in their performance that the more we learn about them, the greater our sense of awe at the wonder of it all.

The authors do not claim to have written an advanced text-book on vision. On the contrary, as they explain in their Preface, the book is intended as an introductory text to whet the appetite of the reader to study the subject more deeply. In my judgement they have been very successful in achieving their aim, since they have covered many aspects of light and colour perception without getting bogged down in too much detail. I can only assure those readers who are inspired to take up the subject in greater depth that they will be setting out on a road that will bring them lasting interest, enlightenment and pleasure.

W. D. Wright
Emeritus Professor of Applied Optics,
Imperial College of Science and Technology,
South Kensington, London

Preface

This book is about some aspects of visual perception, which is how the brain interprets the information sent to it by the eye. The eye is the most incredible and wonderful organ with an immensely complicated structure, and hence a very complicated performance. It can detect brightness brightness differences, resolve small detail, see colour, form, shape, size texture, movement, flicker, and present them to the individual's brain in such a way that he senses, and then 'perceives', or actually consciously experiences the environment—this extends from that immediately around him to the distant stars. Consequently much of this book is concerned with explaining the relationship between the physical stimulus and the resulting psychological perception. This field of study is commonly called psychophysics.

The sheer complexity of the eye has led to much research both from the structural or anatomical side, and also from the operational or physiological angle. Many eminent scientists have taken part in this work. Fortunately it is very well documented, and the study of vision culminated when Helmholtz published his *Physiological Optics* in 1866. Most visual phenomena are to be found in this great work, and although we have since added some details, the immensity of the original achievement still astounds us.

Over the last fifteen years the electron microscope with its enormous power of magnification has revealed incredible details in the retina and other structures of the eye. Although we now know much more, the new —almost overwhelming—details which are shown are much more complex than was ever dreamed of before. This means that unless a considerable amount of research is embarked upon, the problems may multiply at a greater rate than the solutions.

The wealth of learned books, and huge quantities of modern research papers are very daunting to the new-comer to the study of vision. It is for this reason that this book has been written to introduce some aspects of the subject to the general reader and to those working in bordering disciplines. It is hoped that it may be useful to the sixth-form student who wishes to widen his horizons. It might help him to decide whether the subject is one to which he might like to devote further attention in his future studies. The first-year university student of physics, electrical engineering, biological sciences, psychology, the building sciences, architecture, or ophthalmic optics might find this book to be useful additional reading.

In order not to treat the subject in a too superficial manner, and to make it of manageable size, it has been limited to the two basic, but fascinating, visual perceptions of light and colour. Some description is given of the eye and its anatomy and physiology. Since seeing is mediated by the perception of differences of contrast or colour, or both contrast and colour, the perception of brightness is discussed at some length. The remainder of the book is concerned with colour, the nature of colour, how it is produced, specified and measured. The perception of colour is then discussed together with a description of various colour phenomena, and how colour comes into our lives. Abnormal colour vision and its implications is also touched on.

The subject is one which lends itself to beautiful and elegant experiments and demonstrations, and some hints are included on how to perform these inexpensively where this would enhance understanding. Many excellent student projects on vision have been carried out in schools and colleges, and perhaps this book may inspire other students, and possibly suggest new projects. A short bibliography of reasonably accessible literature is given for those who feel that they cannot leave the subject here.

Grateful thanks are due to our colleagues and friends too numerous to mention individually who have kindly given us help and encouragement on this project. In particular we should like to thank Miss J. Upton and Mr C. Bishop for their help in preparing photographs.

We wish to thank our colleagues Mrs J. Birch for Plate 10 (*bottom*), Dr G. L. Ruskell for Plates 1 and 2 (*top*), and Mr S. Harry of the Bradford College of Art and Technology for Plate 9 (*top*). Plate 10 (*top*) is reproduced from *An Introduction to Color* by Ralph Evans (1948) by kind permission of John Wiley and Sons, Inc. and Plate 7 (*top*) is reproduced by kind permission of Munsell Colour Division. Grateful thanks are due to Mr G. Gough of the London College of Printing for his help with Chapter 5.

We would like to record our grateful appreciation to Professor R. J. Fletcher for his support and for his infectious enthusiasm for all aspects of Physiological Optics, and to Professor W. D. Wright for his inspiration and kind help over many years. Lastly we wish to acknowledge the kind help and forbearance of our wives.

C.A.P.
J.E.S.

Contents

1
Introduction

Living as distinct from just existing, is a restless process from the cradle to the grave of receiving information from one's environment, processing it, and deciding whether to act upon it, to ignore it, or to store it for future use. The central processing unit is the brain. The human brain is the size of two clenched fists, weighs 1300 to 1400 g, and is covered by a thin sheet of grey matter called the cerebral cortex, which is spread over the two brain hemispheres. It is wrinkled and folded, rather like a walnut, in order to pack it in, since its area is about 1500 cm^2. This grey matter contains about 10 000 million nerve cells or neurons, and it is when these receive messages from nerve fibres that sensation and conscious experience occurs.

The human body is composed of cells, but some of these, the sensory cells, are specially made to gather information from the environment and send it to the brain. The brain exists in silence and darkness, and cannot by itself hear or see, or experience any other sensation. There are in fact eight different senses, namely sight, hearing, smell, taste, touch, temperature, muscular and orientation sense, and pain. The pieces of information from the outside world of which the organism needs knowledge are the stimuli, and these are either physical or chemical in nature. Thus light is electro-magnetic wave energy; sound is compression wave energy in air; smell and taste are activated chemically; and the remainder (except pain) are physical. Pain can be either physical or chemical in origin.

The sensory cells receive the stimulus energy (that is, they are stimulated) and then transpose it and its variations into nerve signals, which are transmitted to the brain along nerve fibres. In modern terminology one speaks of the sensory cells as transducers since they transpose the stimulus energy into another form—electrical impulses—to which the brain can respond. Exactly how they do this is still not known. Each sensory cell is therefore either a physicist or a chemist, and is also a highly specialized being who can only deal with certain specific stimuli. Thus a light-sensitive cell cannot respond to sound, pressure or chemical stimulation, and a touch receptor cannot respond to light or sound. How they develop this specificity is still a mystery.

It is an interesting point that the nerve fibres and nerve impulses lead-ing from the sensory cells to the brain are similar for all the senses, the only difference is that they terminate in different parts of the cortex of the brain. Therefore the fibres which carry signals to the visual areas at the back of the brain produce the sensation of light, and those carrying similar signals to the auditory regions give rise to the sensation of sound. The brain is however responsible for a much richer experience than pure sensation. Sensation is the crude experience such as light, sound, heat, etc. but the brain activity also results in perception, which is the interpretation of the sensory input in the light of experience. This is the conscious experience of the environment. Thus the little upside down image in the eye is perceived as a chair the right way up some distance in front of the eye, and a larger but similar image appears as a chair nearer to the eye. If we prick a finger the muscular senses tell the brain whether the finger is held at our side or above our head, because the brain perceives the pain as being in the finger and not in itself. Perception is therefore very complex, but it is further complicated by the emotions. A memory store is of course essential, since it is by its use that we quickly and easily interpret the retinal images, significance of sounds and smells and many other sensations and their combinations.

We know therefore that the brain is the seat of our perceptions, but in reality we know very little about it, or how it transforms nerve impulses into sensations, or how it perceives or how it stores information. It is still largely shrouded in mystery. However, it is the size and complex organization of the human brain which gives us our ascendancy over the rest of the animal kingdom. Undoubtedly the development of sight and language have contributed mainly to this dominance. Most of the time the proper operation of the brain is entirely dependent upon being fed with new information. The two major collectors of the information are the eye and the ear. Much of the important information comes from them, and indeed most of this comes through the eye.

It is not easy perhaps for a modern scientist, with such elaborate tools as automated instrumentation and computers, to realise that the human senses still play an essential part in the reception of this information. Nowadays in many experiments, instruments such as photo-electric cells and microphones are substituted for the human eye and ear respectively, but they are only (often poor) substitutes, and their performance must be carefully matched to that of the corresponding human organ. Therefore readings from instruments using such devices cannot be more accurate than those from the eye or ear alone, even though they may be more precise. They are usually obtained more easily and quickly, however. Obviously it is necessary to know a great deal about how both the eye and the ear react in different situations. Furthermore the results still have to be read by a scientist from a meter or from a computer print-out, added to which he must use his judgment in the interpretation of them. In doing this he must know the possible errors and limitations of his instruments, and of course, of himself.

The human brain has now evolved to such a state of sophistication that it is very much more than merely an adequate instrument just concerned with the basic needs of survival and reproduction. The result is the aesthetic spin-off of sport, art, literature, music and so on which so enriches our modern civilized life. Furthermore it allows us largely to control our environment, with all the attendant advantages. However, perhaps the most remarkable of human achievements are in the realms of greatly extending the range of our senses. We can now detect, for example, radio waves, which means we are now aware of the existence of many radio stars which were totally unsuspected 40 years ago; thus we are receiving more information about our environment. It leads one to wonder what the world is really like, since we can only know that part which we are able to detect either directly with our senses or with our available instruments.

Although perception has been studied for a long time, we still know very little about its true nature or its basic origin. It may be we shall never know much more, but research will go on, since man is nothing if not infinitely inquisitive. Indeed Sir Isaac Newton himself said just before he died:

> I do not know what I may appear to the world, but to myself I seem to have been only like a boy playing on the seashore and diverting myself in now and then finding a smoother pebble or a prettier shell than ordinary, whilst the great ocean of truth lay undiscovered before me.

2
Eye and brain

2.1 INTRODUCTION

The first stage in the visual process occurs in the eyeball and it is here that the properties of vision are frequently likened to the operation of a camera although this comparison is rather an oversimplification. A variable aperture called the iris determines the amount of light which enters the eye. The cornea, the ocular media and an elastic lens act as a high-powered variable focusing unit by which the light emitted by an external scene is imaged on a photo-sensitive surface called the retina. It is at the retina and at the higher centres of the brain that information on colour, size, movement and other spatial and temporal properties of the image are extracted by means of chemical and electrical analyzers and finally synthesized with other sensory information stored previously and simultaneously to produce a visual perception. In this chapter we shall discuss briefly some of the properties of the eye and brain which contribute to the present state of knowledge of the perception of light and colour.

2.2 A DESCRIPTION OF THE HUMAN EYEBALL

A horizontal section of the human eyeball is shown in Fig. 2.1. The cornea is a transparent membrane which covers the nearly spherical protuberance at the front of the eye and extends over approximately one-sixth of the surface of the eye. At the limbus this membrane continues as the sclera (the white of the eye) which forms the remainder of the eye's outer coat. The aqueous humour, which is 99 per cent water plus salts and proteins, fills the space between the lens and the cornea. The major volume of the eye behind the lens is filled with a transparent gelatinous mass, the vitreous humour, which also consists mainly of water. No part of the eye is rigid and the shape, which approximates to a sphere with a radius of curvature of approximately 12 mm is maintained by the pressure of the internal fluids.

2.2.1 The iris and the pupil

The iris is a coloured fibrous structure whose central aperture forms the pupil of the eye. Two involuntary muscles, the sphincter and the dilator

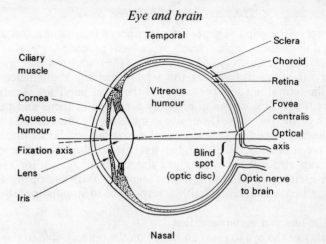

Fig. 2.1 A horizontal section of the right human eye.

control contraction and dilation of the pupil respectively. The pupil
diameter can vary from 2 mm to 8 mm, thereby controlling the quantity
of light per unit area which falls onto the retina (the retinal illuminance).
One interesting property of the retina is that for high retinal illuminances
rays of light which enter via the edges of the pupil are less effective in
promoting vision than more central rays. At lower stimulus levels any
such effect is negligible, all quanta of light entering the pupil being
approximately equally effective. This phenomenon was discovered in
1931 by Stiles and Crawford and is an important piece of evidence in
support of the duplicity theory. This theory states that there are two
types of receptor; one determining high retinal illuminance (photopic
vision) which is directionally sensitive to the incidence of light; and one
determining low retinal illuminance (scotopic vision). Because of the
Stiles–Crawford effect the range of photopic retinal illuminance under
the control of the pupil is effectively limited to approximately 10:1
rather than 16:1 as predicted from the variation in pupil area. Illuminance is normally measured in lux* (the light flux in lm^{m-2}) but the unit
used in visual studies is the troland which is defined as the product of
the pupil area in mm^2 and the luminance (photometric brightness) of
the corresponding scene element. Actually the retinal illuminance is
proportional to the product of the pupil area, the luminance and the
transmittance of the ocular media but as we shall see below this last
factor is rarely known with any precision and consequently has to be
ignored.

* See Glossary of Terms p. 180.

The size of the pupil depends on a number of factors, in particular the retinal illuminance, higher levels of which produce a smaller pupil diameter. It is the quantity of light reaching the retina, not that illuminating the iris which determines the size. When the illuminance varies across the retinal image a prediction of the exact pupil size is extremely complex. Pupil size is also affected by a change in focus and the convergence of the two eyes. If one eye is more highly illuminated than the other the two pupils will remain equal in size but will be smaller than if both were stimulated by the higher level. Psychological and emotional factors can also influence the pupil size which increases, for example, with an exciting or interesting visual stimulus. With increasing age the pupil size becomes smaller and less responsive to stimulation.

2.2.2 The lens and accommodation

The ability of the eye to form an image on the retina is due to the curvature of the surfaces which contain the cornea, aqueous, lens and vitreous and to the refractive indices of these four media (which approximate to 1·37, 1·34, 1·42 and 1·34 respectively). The cornea makes the major contribution to the eye's power due to its high curvature and since its front surface, which is normally in contact with the air, is the only interface where an appreciable change of refractive index occurs. However, accommodation, or the ability of the eye to focus on objects at different distances from the eye, is due entirely to the elastic nature of the lens. This has a crystalline, fibrous structure whose shape can be altered by the action of the ciliary muscles (Fig. 2.1). When these muscles are relaxed the curvature of the front surface of the lens decreases and the corresponding decrease in power allows the eye to focus on more distant objects. A similar but much smaller and therefore less important change occurs in the back surface of the lens. An eye in the normal or emmetropic state can image objects at infinity when unaccommodated and objects approximately 15 cm from the cornea when fully accommodated, the total power of the eye varying from approximately 60 to 70 dioptres.* The development of the lens continues throughout life with a consequent increase in size and reduction in its elastic properties. It is not surprising therefore that a young subject may have an amplitude of accommodation of more than ten dioptres while at the age of 60 years this is reduced to an average of approximately two dioptres.

2.2.3 Absorption of light in the ocular media

Before light can be absorbed at the retina to initiate the visual response it must first traverse the ocular media where a variation in the absorbance with wavelength can effectively change the spectral composition of the

---

* See Glossary of Terms p. 180.

stimulus. In some circumstances this can have an important effect on the resultant colour perception.

For the longer wavelengths (above 500 nm) it is likely that the absorption is determined by the water content of the ocular media. Practically all electromagnetic energy above 1400 nm is consequently absorbed by the ocular media, converted to heat energy and therefore does not reach the retina. Some common sources emit a considerable amount of their energy in the infra red (approximately 70 per cent for a tungsten filament lamp), most of this being converted to heat in the ocular media near the front of the eye. In some circumstances this can cause discomfort or even damage to the cornea and lens so that protective glasses which absorb infra-red light must be worn for some industrial tasks such as foundry work where the eye is exposed to high levels of infra-red radiation. At wavelengths shorter than 500 nm, water is transparent to electromagnetic energy but absorption does occur due to the other components of the media. In particular absorption in the cornea and the lens prevents any wavelengths shorter than 300 nm from reaching the retina. Excessive absorption of ultra-violet light by the cornea can cause 'snow blindness' and protective glasses are therefore necessary when working with sources (e.g. in arc welding) which emit strongly in the ultra-violet region. The lens is the principal medium which appreciably affects the spectral distribution of incident visible light, showing steadily increasing absorbance of wavelengths below 500 nm. Consequently the lens appears yellow when illuminated by white light, a phenomenon which increases with ages above about 20 years. This reduction in transmission with age at the shorter wavelengths is due not only to an increase of the absorbance but also to an increase in the scattering of light in the lens which also has a greater effect on the shorter wavelengths.

The central region of the retina containing the fovea is covered with the macular pigment which absorbs light in the blue region of the visible spectrum with a maximum effect at approximately 450 nm. The distribution of the pigment is roughly elliptical with the major axis horizontal and covering 5° to 10° of the visual field. The density of the pigment is relatively low at the foveal or central region. There is considerable variation in the density of this pigment between individuals but it probably does not vary with age. Both the variation of the lens absorbance (with age) and absorbance of the macular pigment (between individuals) must be taken into account when comparing the colour vision of different individuals.

2.3 THE RETINA

Until the sixteenth century it was thought that the lens was the receptor of light but Kepler and others demonstrated that it was simply an optical component which assisted in producing an image of the outside world on the retina which presumably contained the true receptors of light. Schultze in 1866 was able to identify two types of receptor in a variety of animal eyes, the rods and cones, which are now accepted as mediating the human visual response to low and high retinal illuminances respectively.

Light microscopy, and more recently electron microscopy, has shown the structure of the retina to be extremely complex. There are a vast number of cells arranged in layers perpendicular to the surface of the retina which in turn have a large number of interconnections both between layers and within a layer (Fig.2.2) (Plates 1 and 2 (*top*)). It is assumed that light is absorbed by photochemical action at the receptor

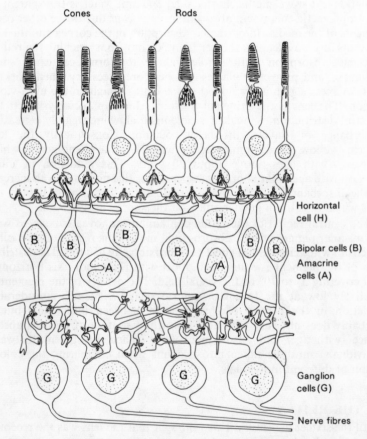

Fig. 2.2 Synaptic connections of the retina—transverse section (after Dowling & Boycott).

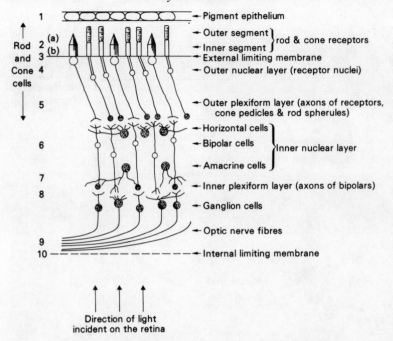

1 — Pigment epithelium

2 (a) — Outer segment ⎫
  (b) — Inner segment ⎬ rod & cone receptors
3 — External limiting membrane
4 — Outer nuclear layer (receptor nuclei)

Rod and Cone cells

5 — Outer plexiform layer (axons of receptors, cone pedicles & rod spherules)

— Horizontal cells ⎫
6 — Bipolar cells ⎬ Inner nuclear layer
— Amacrine cells ⎭

7 — Inner plexiform layer (axons of bipolars)
8

— Ganglion cells

— Optic nerve fibres
9
10 — Internal limiting membrane

Direction of light
incident on the retina

Fig. 2.3 Schematic diagram of a transverse section of the retina showing the ten layers.

cells which in turn activate a neighbouring cell by chemical or electrical changes and that the incidence of the light is thereby signalled from cell to cell through the retinal layers.

In 1852 Müller classified the various layers of the retina, a modified version of which is shown in Fig. 2.3. The most important nerve cells or neurons for present purposes are the receptors (rods or cones), the bipolar and the ganglion cells. In the peripheral retina multiple longitudinal connections of each receptor to a number of bipolar cells, and the lateral connections via the amacrine and horizontal cells effectively mean that each ganglion cell can be excited by a number of receptors, perhaps up to several hundred. This pooling of receptor responses can be used to explain the summation and inhibition effects described in Chapter 3.

The central region of the retina is called the fovea and its subdivisions, the foveola or rod free area and the fixation spot are shown schematically in Fig. 2.4. The horizontal cross section of the region exhibits a central depression which is called the foveal pit. Only cones are found at the base of this depression in the foveola and there are more direct connections from each cone to the bipolar and ganglion cells than are found in

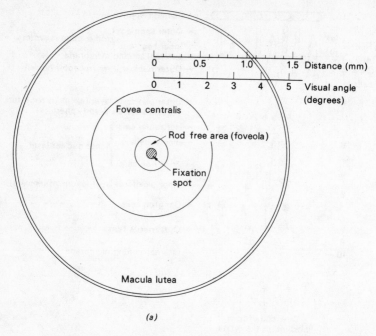

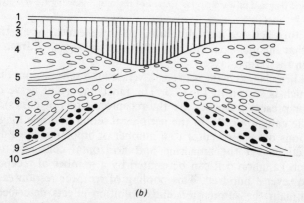

Fig. 2.4 (*a*) Schematic diagram of macular region. (*b*) Schematic section through the human fovea. The numerals relate to the layers of Fig. 2.3.

the peripheral retina. In addition the cones are more closely packed and more elongated than elsewhere in the retina. Thirdly the bipolar and ganglion cells are displaced more towards the edge of the depression. Each of these factors contributes to the special properties of the central fovea. More light quanta can reach the receptors unobstructed by other structures and the corresponding ganglion cells respond only to a small retinal area or pool of receptors. Consequently the central fovea corresponds to the retinal area of maximum visual acuity. This also defines the fixation point, namely the region where an image is formed, by means of eye and head movements when a stimulus detail needs to be seen most clearly.

There are approximately 120 million rods and 7 million cones in the human retina or about 160 000 receptors to each mm^2. Around 25 000 of the cones are situated in the rod-free region of the fovea. Each cone in the central fovea has an outer segment thickness of approximately 2 μm which corresponds to about half a minute of arc of visual space. Elsewhere the thickness of the cones increases by a factor of 3 or more. The rod thickness also varies from 1 to 3 μm. Various estimates have been made of the size of the rod-free area ranging from 0·5° to 1·7° of the visual space. Nevertheless it is safe to assume that the central 1° of the retina is dependent only on cone vision and that the extent of rod intrusion in the central 2° is negligible. Most psychophysical studies of colour vision which is mediated by cones have therefore been restricted to stimuli of this size. A circle of 17·5 mm diameter subtends 1° at the eye when viewed at a distance of 1 m. As Helmholtz suggested, a finger nail viewed at arm's length occupies a field of approximately one degree in diameter.

2.3.1 The photochemistry of the visual pigments

Before light can influence the eye it must be absorbed and this presumably occurs at the retinal receptors which must therefore contain light-sensitive pigments. Of the two types of receptor known to be involved in the visual process a great deal more is known about the photochemistry of the rod pigment than of the cones. Fig. 2.5 shows a simplified diagram of rods and cones. Each has an outer limb where the pigment is distributed on the surface of a series of parallel disc-like structures. The two receptors are distinguished by their shapes, the rods being cylindrical and the cones more conical and shorter. The actual size and shape varies considerably at different sites in the retina, especially among the cones in the rod-free region of the fovea where the density of cones is higher and their shape more elongated and rod like.

The pigment contained in the rod is called rhodopsin and the visual process is initiated by the absorption by one molecule of a single quantum of light. The molecule then changes its structure and bleaching is said to have occurred. In normal circumstances the bleached pigment will return to its original state over a period of time, regeneration having taken place. When all the pigment is regenerated the eye is said to be in the dark adapted state and when bleached by the incidence of light, in

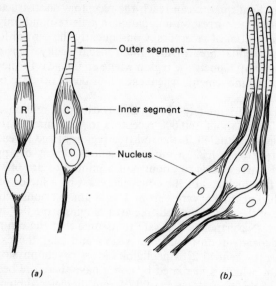

(a)   *(b)*

Fig. 2.5 Schematic diagram of (*a*) peripheral rods (R) and cones (C) and (*b*) foveal cones.

the partially light adapted or fully light adapted state depending on the amount of pigment which is bleached. One important result arising from studies of rhodopsin is that the effect of absorbing one quanta is the same for all wavelengths. Of course quanta of some wavelengths are not as readily absorbed as others as is shown in the absorbance function in Fig. 2.6. We shall see in Chapter 3 that this function is almost identical with the spectral sensitivity, or more precisely the relative luminous efficiency function of the eye, as obtained in subjective or psychophysical experiments, indicating that the absorbance of a quantum is the first stage in obtaining a visual response. In practice only 60 per cent of the light absorbed bleaches rhodopsin, the remainder presumably being converted to heat energy without influencing vision.

In the dark adapted eye rhodopsin, and therefore the rods themselves, appear a reddish colour which changes on bleaching to yellow and finally becomes colourless. Consequently rhodopsin is often referred to as visual purple from the German 'seh purpur' which is more correctly translated as 'visual crimson'. The cones, however, always appear to be colourless when examined by the human eye. If the cones do contain a photosensitive pigment we would not expect them to appear colourless, at least in the unbleached state. It has also been found extremely difficult to extract cone pigments and until recently it has only been possible to conjecture about the existence of cone pigments.

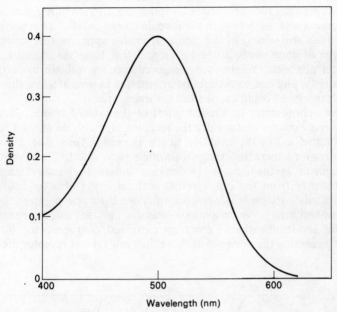

Fig. 2.6 Density spectrum of the rod pigment.

Detection of pigments can now be carried out by three methods: (*a*) by the extraction of the pigment and the measurement of its optical density or absorbance in solution; (*b*) by measuring the absorbance by reflectometry (retinal densitometry) in the intact eye; or (*c*) by spectrophotometry of pieces of retina or even individual receptors. The extraction technique has a number of difficulties associated with it but is extremely useful where the majority of the extract is from a single pigment. This method has therefore been extensively used in studying the absorbance characteristics of rhodopsin but is of restricted use for cone pigments. In the human eye perhaps only 1 per cent of the extract will be of the cone pigment and 99 per cent rhodopsin. Since there is probably more than one type of cone pigment the densitometry of these extracts will be swamped by the rod pigment.

Reflectometry on the intact human eye depends on passing light of a fixed wavelength into the eye through the retina, collecting light reflected after absorption in the retina and comparing the absorbance of the pigment when dark adapted and when bleached to a given level. Many difficulties are involved, not the least being that only one ten-thousandth of the incident light is reflected by the pigment epithelium. However, Rushton and Weale have independently obtained results which distinguish between rods and cones by following the rate of regeneration of the pigment after bleaching in the foveola (cones only) and the peripheral areas (rods and cones) of the retina. It is also apparent that more than one type of cone exists in the retina and that these do indeed contain different pigments. While such measurements are difficult to carry out successfully, and can sometimes be interpreted in several ways, they have the advantage of being carried out on living eyes.

Spectrophotometry of small regions of the excised retina containing 100 or more cells is useful since the receptor pigments do at least remain in the retina unlike the situation in the extraction technique. However, apart from an increase in signal-to-noise ratio due to the removal of the pigment epithelium, the technique suffers the disadvantages of reflectometry from the intact retina without being able to distinguish satisfactorily between the presence of more than one pigment. Microspectrophotometry, where a single beam of light is passed through one receptor and its absorbance spectrum measured, overcomes the difficulty of distinguishing the presence of more than one type of receptor pigment.

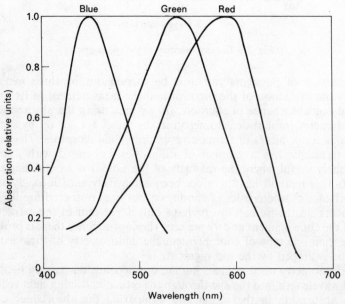

Fig. 2.7 Microspectrophotometric recordings of single cones of human and primate eyes. Each absorption curve has been normalized to give a maximum response of unity. The cones absorbing maximally in the short medium and long wavelengths are normally referred to as blue, green, and red cones although the latter peaks in the orange wavelengths.

Fig. 2.7 shows the absorbance spectrum measured in cones of human and monkey retinae. The fact that these fall into three groups of different spectral sensitivity is strong evidence in support of the trichromacy (three receptor) theory of foveal and cone vision.

We have suggested that there are cone pigments and that these absorb light when mediating photopic vision. We have not, however, explained why the colour of the cone pigments cannot be observed while that of the rods can. Liebman has recently suggested that the energy required to illuminate a single dark adapted cone, suitably magnified so that its colour could be seen, is more than sufficient to bleach the cone in less than a second. A large area of the retina containing mainly rods can be readily obtained and viewed under less illumination for a longer period of time before bleaching is complete and its colour removed. This is unfortunately impractical with cones but this inability to see the pigment colour is not in itself evidence that cones detect colour information by some other means than a photochemical change, although alternative theories have been proposed.

2.3.2 Neural transmission

Fig. 2.8 shows a hypothetical group of nerve cells or neurons. Each neuron is an independent cell containing a nucleus and inward going fibres called dendrites and outward going fibres called axons. By means of these fibres electrical signals are respectively received and transferred to adjacent cells. On receiving a signal the potential in the cell is changed which in general causes electrical pulses to be transmitted along the axon to the junction with a neighbouring cell. These pulses are changes in potential rising to the maximum amplitude and falling to zero within

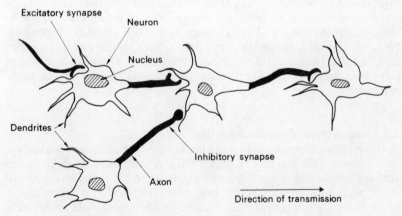

Fig. 2.8 Schematic diagram showing interconnections of neurons.

about a millisecond. The amplitude is always of the same size and the pulses, called action or spike potentials, are therefore said to obey the 'all or none' law. Information on the magnitude of the neuron's response is consequently contained not by the amplitude but by the number of pulses and frequency with which they occur. In general a greater number or higher frequency is indicative of a greater response.

When the action potentials arrive at the junction, or synapse, of a neighbouring neuron a chemical substance called a transmitter passes across the gap at the junction and excites the second neuron, causing it to produce action potentials along its axon. Alternatively the second neuron may be inhibited and any action potentials which are already present will diminish in number (or frequency) or perhaps cease altogether. When neurons are close to each other as in the retina the axon may be omitted and the signal transferred more directly as continuous or 'graded' changes of potential. The individual optic nerve fibre, by means of which the signal is transferred from the eye at the optic nerve or blind spot to the brain, is simply the axon of the final neuron in the retinal pathway, the ganglion cell. Hence the eye, or at least the retina, is considered as part of the brain. Fig. 2.8 simply shows a single chain of cells but it should be remembered, as indicated previously in Fig. 2.2, that the retinal connections are in three dimensions and there may exist return paths by means of which cells in a later layer of the visual pathway may influence the response of earlier cells. In some species, including the human retina, the apparent connections are so complex that many such possibilities exist.

The response of a neuron can take more than one form. It may for example fire or give rise to action potentials when a stimulus is switched on and cease this response when switched off. This is known as an ON response. Alternatively it may show no response or a reduction in firing when the stimulus is on but show a brief series of action potentials after the stimulus is removed. This is known as an OFF response. Frequently both ON and OFF potentials occur for a given stimulus and this is called an ON–OFF response. Information is therefore conveyed in a variety of ways and many of the subtleties of this neural coding remain to be explained.

2.3.3 Neural transmission in the retina

In the vertebrate eye, including the human eye, the retina is inverted, i.e. the receptor cells are at the back of the retina. Incident light travels through the various retinal layers which are transparent and it is absorbed in the outer limbs of the receptor cells where the pigment is concentrated. Most of the light which is not collected by the receptors is absorbed in the black pigment epithelium which therefore reduces the disturbing effects of stray light. The main visual role of this pigment is probably in the assistance of receptor pigment regeneration.

The next stage in the transmission of the visual information is to convey it from the receptors to the bipolar and ganglion cells and hence along the optic nerve to the lateral geniculate body and then to the visual area of the brain (Fig. 2.13). Much of our information at the present time rests on studies of the retina of amphibia and fish, many of whom possess good colour vision (see Chapter 11). It is of course open to question whether the results of such studies can be related to other species such as man. However, where the anatomy is similar we can reasonably expect the general trends to be similar but certainly not identical.

The technique which has produced the most definitive results in tracing transmission involves placing micro-electrodes at various stages in the visual pathway and recording electrical changes both close to the apparent pathways and inside the neurons themselves. Micro-electrodes are fine needle-like electrodes with tips as small as 0·1 μm, constructed either of an insulated metal or a finely drawn glass pipette containing a concentrated conducting fluid.

In the receptors and neurons up to the bipolar cells (Fig. 2.2) a graded potential or a difference in potential across the membrane of the cell can be recorded which varies in amplitude with the level of the stimulus and sometimes with its wavelength. The bipolar cell is to some extent a relay station of information from the receptors although, since several receptors may converge on a single bipolar cell, there exists the probability that information is being collected over quite a large area of the retina, not just from a single receptor. The horizontal cells (Fig. 2.2) can perform two operations. First, they can transmit information from receptors to bipolar cells to which they may have no direct connection. This lateral connection can be of an inhibitory nature so that a bipolar cell may be raised to an excitatory state by stimulation of receptors to which it is connected directly, and to an inhibitory state by more peripheral receptors. Maximum excitation will then occur if only the central receptors are illuminated and the surround remains dark. Secondly, the horizontal cells may feed back from one receptor cell to another, inhibiting or reducing the response which would otherwise be transmitted by the second receptor. Both these types of lateral connection may be partly responsible for the spatial and temporal contrast effects described in Chapters 3 and 8. The bipolar cells drive both the amacrine and ganglion cells, the convergence of several bipolars to a single ganglion and the lateral connections supplied by the amacrine cells providing further possibilities of a pooled response originating from several receptors and mechanisms for further inhibitory effects. It is not until recordings are made in the amacrine and ganglion cells that the signal is transmitted by means of action potentials, as opposed to graded potentials, and these are found again in the optic nerve, the axon of the ganglion cells, and at later stages of the visual pathway.

Micro-electrode recordings have been made at the ganglion cells in a large number of species, including the monkey, where the mechanisms studied are probably similar to those existing in man. Many of these are extracellular recordings which probably reflect the course of the action potentials along individual optic nerve fibres. We have already mentioned that the pooling of receptor electrical responses to a single ganglion cell occurs. This is also supported from anatomical studies of primate retinae which indicate the presence of some 7 million cones, 120 million rods, and less than 1 million nerve fibres. We therefore expect some ganglion cell responses to reflect the excitation of several hundred receptors or more. Thus, it is not surprising to discover ganglion cells which respond when any part of a large area of the retina is stimulated. The extent of this retinal area which when stimulated causes a cell to respond is known as the receptive field of that cell. The spatial and temporal organization of the ganglion receptive fields shows that a great deal of modification of neural signals occurs in the retina, i.e. before the signal is transmitted to the higher centres of the brain. For example, in

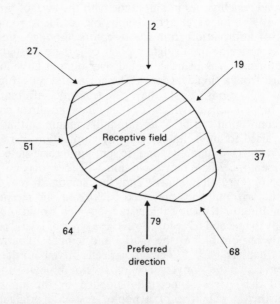

Fig. 2.9 Movement detector at ganglion cell of a rabbit. The hatched area represents the receptive field of the ganglion cell when a small spot of light moves across the retina. The frequency of action potentials is greatest (79/s) when the spot moves in the direction of the bottom arrow.

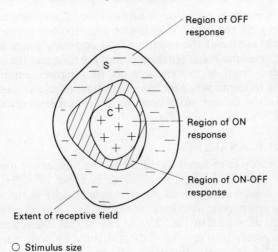

Region of OFF
response

Region of ON
response

Region of ON-OFF
response

Extent of receptive field

○ Stimulus size

Fig. 2.10 Receptive field of a ganglion cell showing an ON centre, OFF surround to a small spot of light.

rabbits approximately circular receptive fields have been discovered which cause a ganglion cell to fire—produce action potentials—only to a small spot of light moving in a particular direction (Fig. 2.9). In the monkey and many other species ganglion cells with concentric antagonistic (centre ON, surround OFF or vice versa) receptive fields have been found. Referring to Fig. 2.10, a small stimulus illuminating the retina in the central (*C*) region will excite the ganglion cell, that is the number of action potentials per second will increase when a light is switched on. If the stimulus is moved to the surround region (*S*) the response will decrease or fire only when the stimulus is switched off. A maximum rate of firing occurs when only the central region is stimulated and this is maximally reduced (or inhibited) when the surround is completely illuminated, the extent of both the excitation due to *C* and the inhibition due to *S* increasing with the illuminance level. Some ganglion cells show the opposite responses, the central region producing inhibition and the peripheral region excitation. Antagonistic colour effects have also been found. For example the central region may cause excitation when stimulated with red light and the surround cause inhibition when illuminated with blue light. If the entire receptive field is illuminated with white light which provides both long wavelengths to the centre and short wavelengths to the surround then the corresponding excitation and inhibition will cancel out and the ganglion cell will continue to fire at the spontaneous rate found for no stimulus. In this arrangement it seems likely that only cones with their maximal spectral response at the longer wavelengths (the red sensitive cones of Fig. 2.7) are providing the response in the centre of the receptive field and only cones with a short wavelength response (blue sensitive cones) are affecting the inhibition in the surround. Both of the above types of ganglion response may be contributing to luminosity and colour contrast effects and seem to form the basis for the further visual analysis at cortical level discussed below in section 2.7, p. 29.

2.4 COLOUR ANALYSIS

We have already introduced the idea that the human eye has four types of receptor, namely a single rod and three cones of differing spectral sensitivity: a blue cone (which absorbs light mainly in the short wavelengths), a green cone (medium wavelengths), and a red cone (long wavelengths) as shown in Fig. 2.7. At low retinal illuminance the eye's response is determined by the rods and since all rods have the same spectral sensitivity two stimuli of different wavelengths will produce identical responses if their levels are suitably adjusted, i.e. the rods do not assist in distinguishing different colours, only variations in luminosity (or subjective brightness). This is not so with high stimulus levels where the cones initiate the visual response. In this case a wavelength at say 450 nm will cause a greater response in a visual channel dominated by the blue cone and a relatively small response in mechanisms activated by the green and red cones. Similarly a wavelength at 650 nm will mainly activate the red cone. Hence, provided that the three cone responses remain separated in the visual pathway, it will be impossible to make the responses to these two wavelengths identical simply by adjusting the

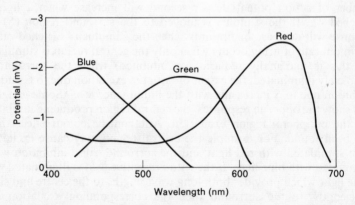

Fig. 2.11 Change in potential in the inner segments of single cones of the carp when stimulated by monochromatic light.

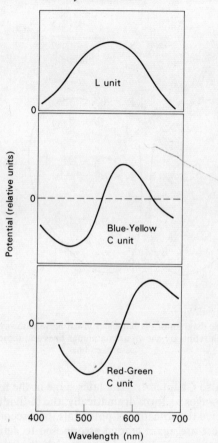

Fig. 2.12 *S*-potentials. The simple *L*- and the two *C*-potentials recorded from horizontal cells in fish.

level of the stimulus. We therefore possess, by means of the cones, the ability to discriminate both changes in the level and colour of a stimulus.

Studies of the electrical properties of the retinae of fish and other species which have good colour vision have shown that the individual characteristics of the three cones do remain distinct. Micro-electrode recordings in individual cones show that the cone spectral sensitivity functions are immediately transferred to the electrical responses. Fig. 2.11 shows electrical recordings in the individual cones of the carp where, as expected with fish, the maximum sensitivity of each type of cone is shifted to longer wavelengths than that obtained for humans

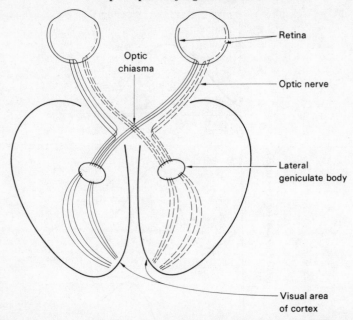

Retina

Optic chiasma

Optic nerve

Lateral geniculate body

Visual area of cortex

Fig. 2.13 Schematic diagram of the binocular visual nerve pathways. Nerve fibres leaving the left side of each retina are shown as continuous lines and those leaving the right side as broken lines.

(Fig. 2.7) (see also Chapter 4). At a later stage in the fish retinal pathway the response changes its form dramatically, the individual cone responses combining to give the so-called S-potentials, first recorded by Svaetichin (Fig. 2.12). These are again graded as opposed to action potentials. No colour information is lost, the L-potential being a summation of the three individual cone responses and therefore ideal for transmitting information of the stimulus level, and the two C-potentials still distinguish between the eye's response to long, medium and short wavelength stimuli and are therefore ideal for transmitting information relating to colour and colour differences. In the monkey similar types of response (often called opponent colour responses) have been recorded in the lateral geniculate body. These colour responses are discussed further in Chapters 4 and 10.

2.5 BINOCULAR VISION

The advantages of having two eyes can be twofold, either to extend the field of view or to provide two separate retinal images of the same scene as observed from two locations. Many animals have their eyes placed laterally to obtain the former advantage, thereby possessing panoramic vision. Man, the other primates, and certain other animals, have their eyes displaced in the same plane to produce binocular vision, i.e. the two visual fields overlap and in normal vision these are fused by the brain to produce a single but now three-dimensional view of the world. The horizontal separation of the eyes results in differences between the two retinal images which can be interpreted by the brain to distinguish between the position or depth of different objects in space.

The anatomy of the visual nerve pathways is shown schematically in Fig. 2.13. The optic nerve fibres linked to different parts of the retina cross at the optic chiasma, the left half of each retina feeding information

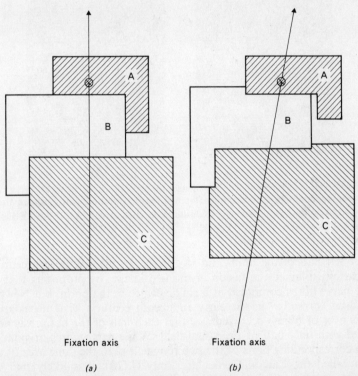

Fig. 2.14 Conflict of depth perception arising from monocular clues alone. The width of the three cards decreases in the order A, B and C. Card A is nearest to the subject, then B, and C is furthest away. When viewed with one eye along fixation axis (*a*) card A will appear to be furthest from the eye due to the smaller retinal image. When viewed along axis (*b*) the overlapping of the cards assists the subject to see the cards in the correct order. Very careful positioning of the cards and the avoidance of shadows are required to experience this illusion.

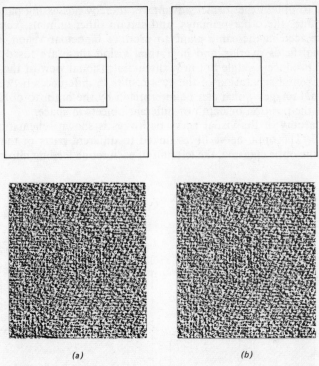

(a) *(b)*

Fig. 2.15 Stereoscopic images. The two images (a) and (b) are presented to the left and right eyes separately. In both examples (top and lower figures) the central square area is shifted 2 mm to the right in (a) and left in (b). When fused with a stereoscope the central square will appear to float in front of the background.

to the left half of the brain and vice versa. In the seventeenth century Descartes explained the fusion of the two images by proposing that all optic nerve fibres combined at a single centre in the brain, but Newton proposed correctly—in the early eighteenth century—that decussation (cross over of fibres) occurred and that each half of the brain was concerned with only half of the visual field. While nerve fibres originating from corresponding parts of the two retinae lead to the same area of the brain called the Lateral Geniculate Body (LGB), it is likely that the two responses are not combined until the cortex. Electrical recordings, using micro-electrodes, show that single neurons in the visual region of the cortex are frequently driven by stimuli from both retinae. At the LGN each neuron seems to be driven by only one retina.

It should not be supposed that binocular vision is essential for depth perception since many clues to the position of an object in space are available from a single two-dimensional retinal image. For example the visual angles subtended by objects of known size, the relative apparent movement of objects at different distances when the eye is moved (parallax), the required degree of accommodation, the overlapping of one retinal image by another, and the fact that more distant stimuli appear blue due to scattering in the atmosphere are all useful clues for determining distance visually. (Sometimes the clues can be contradictory and amusing, but highly informative perceptual phenomena can occur as shown in Fig. 2.14.) Nevertheless our ability to judge depth is greatly enhanced by stereopsis or the fusion of two images of the same scene.

It was once thought that contours were essential to produce stereopsis but Fig. 2.15 shows that this is not so. In the upper part of this figure the left and right fields of the highly-contoured diagram of a small square on a background will, when fused by the left and right eyes, give the impression that the smaller square stands out from the background. In the lower figure a corresponding area will produce a similar effect. Here the apparent random arrangement of dots are identical except that a small square of dots in the left half of the figure has been displaced by 2 mm to the right as compared with the other half of the figure. No contours and therefore no impression of an inner square region are seen in either pattern when viewed separately but when fused using a stereoscope the inner square again appears to stand in front of the background. Such random patterns are called Julez patterns after their originator who has studied many related visual effects during recent years.

2.6 MOVEMENTS OF THE EYE

As shown in Fig. 2.16 there are 6 muscles which control the movement of the eye. By contraction and relaxation the eye can be moved up or down, to the left or right and also be rotated so that the eyes tend to remain horizontal as the head is tilted. Any eye movement can be considered to be made up of one or more of these three types of rotational movements. The action of each muscle for any movement is complex and will not be considered further although it can be noted that any movement involves the contraction or relaxation of at least 4 muscles. Binocular movement of course involves the operation of up to 12 muscles, the contraction–relaxation roles of some being reversed in, for example, a simple movement of both eyes to the left as opposed to a convergence of both eyes towards a near object.

The eye is rarely, if ever, stationary, small involuntary movements being made even when an attempt is made to fixate an object. Fig. 2.17 shows the locus of a typical series of eye movements during an attempt at steady fixation. Three types of movement are seen. First, there exists a small tremor of low frequency (approximately 50 Hz); secondly flicks

Fig. 2.16 The right eye viewed from above showing the 6 muscles which control the eye position.

or saccades (normally corresponding to a few minutes of arc) which may occur once or twice a second with angular velocities of hundreds of degrees per second; and thirdly a slow 'drift' occurring between saccades.

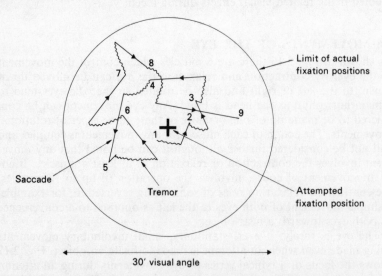

Fig. 2.17 Eye movements during an attempt to fixate an object. Actual fixation positions are indicated by the numbers 1 to 9.

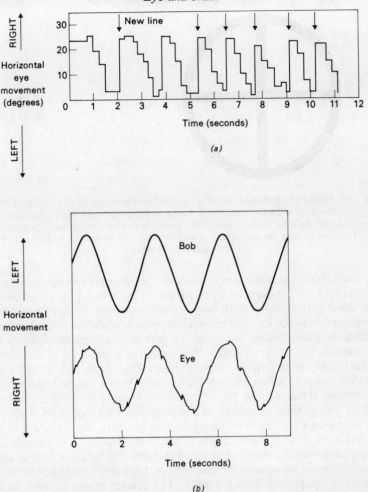

Fig. 2.18 The horizontal component of eye movements when (*a*) reading and (*b*) attempting to follow the bob of a moving pendulum.

Each of these movements occurs when voluntary eye movements are made such as when reading (see Fig. 2.18). The figure also shows the horizontal eye movements of a subject attempting to follow the oscillations of a simple pendulum. In general the expected sinusoidal movement is exhibited but corrective and involuntary saccades are also apparent.

Two features of eye movements have interested visual psychophysicists in recent years. First, there is to some extent a loss of vision during an

Fig. 2.19 Fading of a stabilized retinal image produced by means of an after image. If the centre of the cross is fixated steadily for 1 min. and then the right hand spot fixated for several seconds an after image is seen. This exhibits both total and partial fading (perhaps only the cross or only the outside circle will be seen). The image can be restored by eye movements or blinking.

eye movement, the minimum amount of the stimulus retinal illuminance required for it to be perceived (the threshold level) being increased during this short period which may only last about 20 ms. A more interesting phenomenon is the loss of vision which occurs when the retinal image is artificially stabilized on the retina so that even during involuntary eye movements the image remains on exactly the same area of the retina. In this case the perception of form and colour greys out within a few seconds and only a dim uniform field is seen. Vision can be restored only by moving the image on the retina or by causing some change in the retinal illuminance or colour of the stimulus. One can infer from this phenomenon that the involuntary eye movements, the tremor and the saccades are an important feature of the visual process, it is possible that they play a role in enhancing contrast and limiting fatigue effects at some stage in the visual process. Sometimes, perhaps when stabilization is not perfect, only partial fading occurs. The reader might be able to experience this phenomenon by fixating as steadily as possible for about a minute the centre of the cross in the circle of Fig. 2.19. If the small black spot is now fixated an after-image (see Chapter 9) will be seen. This after-image is stationary on the retina and can be seen to fade and reappear within a few seconds. Occasionally only the cross or the circle will fade, perhaps indicating that the perception of one feature of a stimulus is stored separately from another in the brain.

Another interesting phenomenon which the reader might like to observe is the Purkinje tree. This is the appearance of the blood vessels and capillaries which are always present at the back of the eye (fundus). These are very close to the receptor cells and form sharp shadows on most regions of the retina. Why then are these not seen when we look at a uniform field such as a piece of white paper? To a good approximation these shadows always fall on the same region of the retina and can be considered as an excellent example of a stabilized retinal image. However, it is possible to illuminate the eye from an unusual angle and momentarily at least to see this rather beautiful phenomenon. A small torch bulb is placed close to one side of the eye—or above or beneath the eye—so that the retina is illuminated by oblique light or preferably by diffuse light through the sclera. If the subject then views a wall in an otherwise dark room and moves the torch to and fro (just a few millimetres) the Purkinje tree can be seen for fairly long periods (Fig. 2.20). As with many visual phenomena it is easier to perceive after having seen it once before.

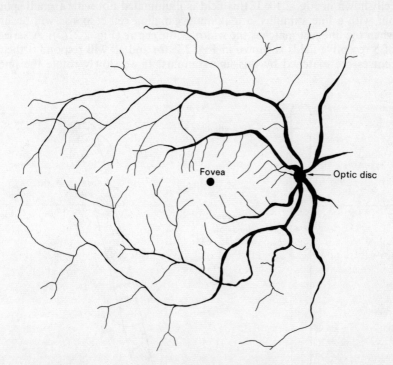

Fig. 2.20 The Purkinje tree caused by the image of the capillaries across the inner face of the retina.

2.7 FEATURE DETECTORS AT CORTICAL LEVEL

We have seen how at retinal level a ganglion cell does not just transmit information on the level of the stimulus but that its response may also reflect more subtle characteristics of the stimulus such as the relative illuminance over different parts of the retina. During the last fifteen years considerable advances have been made in our knowledge of the manner in which features of a stimulus are analyzed in the visual pathway. Prior to this it had been established from studies of the effects of war wounds which had caused lesions in local areas of the brain, which cortical areas were responsible for transmission of certain sensory and motor events. We will restrict ourselves here to a brief discussion of the type of experiments carried out by Hubel and Wiesel in the last decade on cats and monkeys and to the simpler aspects of a model of the analysis of visual information in the retina, Lateral Geniculate Body (LGB) and the area of the cortex known as area 17 or the visual cortex.

Consider again the centre-surround receptive field of a retinal ganglion cell shown in Fig. 2.10. If this field is illuminated not with a small spot but with a line stimulus, a maximum ganglion cell response will occur when the line just matches the width of the centre (Fig. 2.21(*a*)). A series of 5 receptive fields is shown in Fig. 2.21(*b*) and all will respond if their centres are matched by this line stimulus. If we slowly rotate the line

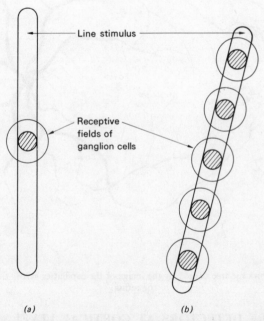

(a)   (b)

Fig. 2.21 Receptive fields and the detection of orientation. (*a*) Line stimulus positioned to produce an optimum response from a ganglion cell. (*b*) Line stimulus positioned to produce an optimum combined response from five neighbouring receptive fields.

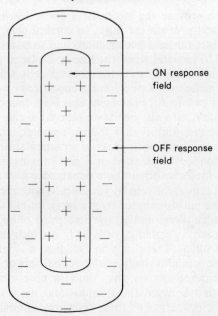

ON response field

OFF response field

Fig. 2.22 Receptive field of a simple cortical cell. A small spot of light illuminating (*a*) the central area (+) produces an ON response and (*b*) the immediate surround produces an OFF response. A maximum ON response is obtained when a line stimulus of the correct size and orientation illuminates the central area only.

about the middle of the 5 fields the ganglion cell corresponding to this field will continue to fire as before but the other 4 ganglion cells will respond less and less as more of the stimulus falls onto the inhibitive surround. If the rotation is continued eventually only the cell corresponding to the middle receptive field will continue to fire. We have then an effective line detector in which the total response of the 5 ganglion cells is maximal when the line is in a certain position on the retina of correct width and orientation. Hubel and Wiesel found the antagonistic centre-surround field at both retinal ganglion cells and also in the lateral geniculate nucleus although the spatial analysis appears to be a little 'sharper' in the lateral geniculate nucleus. The latter should not therefore be dismissed just as a simple relay station. However, in the cortex they discovered a new type of receptive field, as shown in Fig. 2.22, which they called a 'simple' cell. One such cell gave its maximum response to a line stimulus of fixed width and orientation. It is reasonable to suppose that a number of ganglion or lateral geniculate nucleus cells have the centre of their receptive fields distributed along this line, as suggested above, and that all of these provide the input for a cortical cell. Other groups of ganglion cells will provide the input to cortical cells responding maximally to dark lines on light backgrounds and to edges. Whereas each of these cells fails to respond when the line is displaced from the receptive field or if the orientation is changed, other cells will respond if their receptive fields match the new stimulus conditions.

Other groups of cortical cells have also been found which appear to collect the output from a number of simple cells having similar properties. For example Hubel and Wiesel found 'complex' cortical cells which responded only to stimuli with edges of a particular orientation but now these cells responded to an edge or a line moving over large areas of the retina provided the orientation remained constant. It seems quite likely that such cells are driven by simple cells with receptive fields of the same orientation but distributed over a large retinal area.

It seems possible therefore that the perception of an object is based on the output of a large number of cells each of which reports on one property of the stimulus. These may in fact feed a single cell which only responds to one stimulus such as the retinal image of a human face. However, we cannot assume that there is necessarily such a simple causal relationship between the stimulus and the response. It is quite possible that the properties of these cells are continually changing and are influenced by other factors than the immediate stimulus. Cell properties may be shaped by previous experience, memory, expectation and even personal likes and dislikes. It is by no means certain that human perception depends to any large extent on the computer-like process revealed in the experiments discussed above although they undoubtedly illustrate an important feature of the initial stages of the visual process.

2.8 HUMAN EVOKED POTENTIALS

Studies of the electrical process in the visual pathway using for example micro-electrodes are restricted to animals, the corresponding human process being deduced by analogy and extrapolation only when the anatomy and behaviour of the animal and human are sufficiently similar. However, support for these deductions can be gained from studying electrical responses from the eye and brain by attaching surface electrodes near to associated visual areas of the human subject.

One such potential is the gross response of the retina which is described by the electroretinogram (ERG). When the eye is stimulated a change in potential occurs across the retina which contains characteristics of the stimulus and the retinal processes involved. A typical response to a brief flash of light is shown in Fig. 2.23. Immediately after the onset of the stimulus there is a small change in potential which has been shown to reflect the properties of the receptors. This so-called early receptor

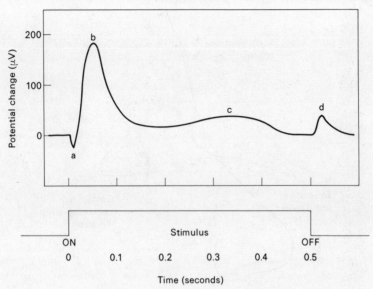

Fig. 2.23 A schematic diagram of the ERG.

potential (ERP) is in fact too small to be seen in Fig. 2.23. This is followed by the late receptor potential or *a*-wave again related to the receptor properties. Next there occurs the *b*-wave which is probably dependent on the state of the bipolar cells and other neurons in the corresponding region of the retina. The slow rise in potential (the *c*-wave) following this is dependent on the pigment epithelium. An OFF effect or *d*-wave may be seen as the stimulus is extinguished. This ERG can be recorded with human subjects by placing an electrode near the cornea either by means of a special contact lens or a saline soaked wick electrode (Fig. 2.24). The indifferent or reference electrode can be attached to the forehead, the ear or some other neutral area. Clinically such recordings have been used to test the physiological state of the retina either prior to or after treatment of visual disorders. The ERG can also be used to complement psychophysical studies since it reflects to some extent the first stage of the visual process in an objective manner, i.e. it does not require introspective statements or actions from the subject.

The electroencephalogram (EEG) or 'brain wave' is recorded by means of surface electrodes in the occipital area of the scalp near the back of the head (Figs 2.24 and 2.25). This has been found to contain information related to a visual stimulus. On the assumption that this is reflecting the gross response of the brain nerve cells, it can be used to

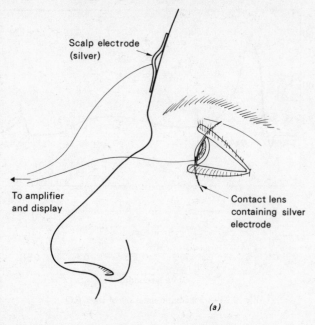

Scalp electrode
(silver)

To amplifier
and display

Contact lens
containing silver
electrode

(a)

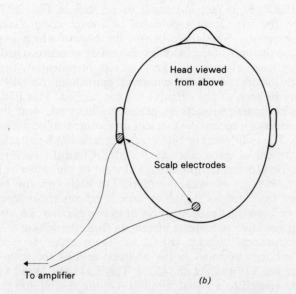

Head viewed
from above

Scalp electrodes

To amplifier

(b)

Fig. 2.24 Methods of recording human ERGs (*a*) and EEGs (*b*).

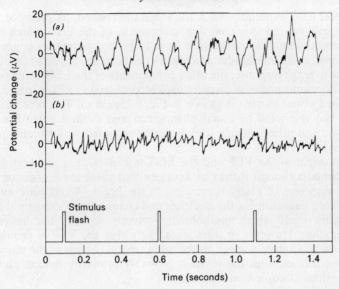

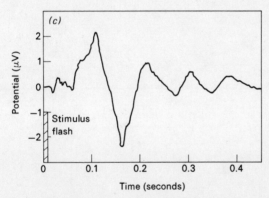

Fig. 2.25 The human EEG (*a*) with the subject relaxed (*b*) with a repeated flashing light and (*c*) when the response following stimulus (*b*) is averaged.

monitor the state of the visual response on reaching the cortical area. As with the ERG it has been used as an objective measure in both clinical and psychophysical studies. Fig. 2.25(*a*) shows a human EEG recorded from a subject relaxed and with his eyes closed. The most apparent feature is the near sinusoidal potential with a frequency of approximately 10 Hz known as the alpha rhythm. While this is a very common feature of the occipital EEG especially when the subject is relaxed, its presence is not necessarily related to *vision*. Any component of the EEG which relates to a visual stimulus change (the visually evoked potential) is generally very small (with amplitudes of a few μV) but can be recovered by observing the average response, the other components of the EEG being treated as background noise. A visually evoked potential (VEP) in response to a flashed visual pattern is shown in Fig. 2.25(*b* & *c*). This type of analysis is not restricted to visual phenomena and evoked potentials from acoustic and other sensory stimuli can be recorded with suitably placed electrodes.

The origin of the VEP and the EEG in general, has not been finally established although it may be assumed that these are a gross or combined response of many nerve cells in the brain. A complete analysis will have to account for the electrical and chemical transmission through the scalp itself which undoubtedly severely distorts the underlying information. However, it does seem likely that the VEP is primarily a response to stimuli incident on the more central regions of the retina and because this includes the fovea it probably primarily reflects photopic rather than scotopic visual conditions.

3

Seeing brightness

3.1 LUMINOSITY

If we find ourselves during the day enveloped in a very dense fog or mist, when quite literally we cannot see our hand in front of our face, we see just light without shape or form. This is simple light sensation, and the corresponding simple perception is one of apparent brightness. Nowadays we use the term luminosity for this subjective quality to avoid confusion with the physical brightness which we now call luminance. This elementary sensation could of course be mediated through an extremely simple 'eye' which could be just one light receiver somewhere on the surface of the body. But very little information is transmitted by unmodified luminosity in this type of situation. If we were to simulate this condition by placing the observer in the centre of a uniformly illuminated spherical room with white walls, we could switch the light on and off, and convey a limited amount of information. This is virtually what is done when an Aldis lamp is used to signal Morse Code. It is a useful but tediously slow way of transmitting and receiving information, and information transfer is of course really what seeing is all about. We could also colour the light (being humans with a most complicated colour receiving mechanism), but apart from this—and switching on and off—little information can be conveyed.

In fact the eye is able to detect not only light but contrast, and this increases enormously the potential as regards the amount of information which can be transferred. A small object can only be seen when super-imposed on a larger one if the two differ in luminosity or in colour, or differ both in luminosity and colour. These differences are known as contrast; differences in luminosity being called luminosity or brightness contrast and differences in colour, colour contrast. Many authors also use the term contrast in an objective sense, for example, to refer to the relative difference in luminance between two adjacent patches of light. This can be confusing to the reader and therefore we will restrict our use of this term to indicate a subjective difference except where this is prefixed by an objective term as in luminance contrast.

In general increasing the amount of light reflected from a surface increases the contrast. This is shown in Plate 2 (*bottom*). There are, however, situations in which the eye plays tricks and such a case is shown in Plate 3 (*top*). The part of the grey ring surrounded by black has a higher luminosity than the part surrounded by white, even though each half of the ring reflects the same amount of light into the eye. There appears to be an in-built mechanism in the eye for increasing the perceived contrast. The mechanism by which the contrast is increased is thought to be due partly to lateral enhancement and inhibition mediated by the cross connections between neurons in the retina which can influence the perception in adjacent areas (section 2.3, p. 8). It is probably also due to differences of interpretation in the brain.

3.2 THRESHOLDS

As in most human activities it is helpful to know the limits of performance. If these refer to athletes, for example, the high jump or time taken to run a mile, the maximum or minimum values are written down as records. In the case of our senses, the best performance is usually called a threshold. So the intensity of the faintest light which can be seen on average in a particular situation is called the threshold value. It is quite reasonable to suppose that if you were presented with a light slightly above the threshold intensity, you would see it, and if presented with one slightly below you would not see it. Unfortunately it is not as simple as this, and you might sometimes see it even if it were considerably fainter, and on the other hand you might not see it even if considerably above this value. Why is this? There are two reasons. First, light itself is not emitted continuously from a source, but is fired off in little packets called 'quanta' which are shot out in a random way. Although for a steady light source, the average number of quanta emitted in a given time is constant, it does not follow that precisely the same number are emitted in two equal successive intervals of time. An analogy is for example in a rainstorm, the number of raindrops falling in a second on two adjacent square inches of soil may be slightly different, although the number falling on each patch would be fairly precisely the same, if the counting were extended over a few minutes. Consequently the human threshold will vary because of these random variations in the stimulus which occur no matter how exactly we attempt to control the physical conditions.

The second factor is that the rods and cones which receive the quanta are not always in the same state of sensitivity. For instance, if two quanta of light fall on a pigment molecule in a rod in very quick succession, the second quantum may have no visual effect, because the rod molecule will not have had time to recover its sensitivity in the short time between the two impacts. Also two successive quanta may fall on different parts of the retina which have slightly different sensitivities. Of course some quanta are reflected back from the transparent surfaces of the eye, and some are absorbed by the media of the eye before they reach the retina. Some again are not absorbed by a rod or cone even if they strike it head-on. For these reasons the emission and absorption of light and the stimulation of a light receiver element in the eye are to some extent governed by the laws of chance.

It may perhaps come as a surprise to some, that scientists are quite used to dealing with things which happen by chance. Many physical phenomena such as heat, gas pressure, atomic fission, etc. are chance phenomena, and effects are computed for the average behaviour of perhaps hundreds of thousands of atoms, without knowing precisely what is happening to each one individually. The mathematics involved is usually complicated, and of course goes under the general heading of statistics. These factors are important when considering visual thresholds because we are often dealing with the absorption of a very small number of quanta.

Thus if we are doing an experiment on the visual threshold we must present a number of patches of light to an observer in succession. If we alter the luminance of the different presentations in a random form, we can ask the observer each time to state 'yes' or 'no' depending upon whether he sees it or not. (This is called the method of constant stimuli.) We can now draw a graph of the percentage of 'yes' answers plotted against the actual luminance of the stimulus (Fig. 3.1). This shows that at very high luminances the light is always seen, and at very low luminances it is never seen. But in between it is seen on a proportion of the times it is presented. The threshold value of luminance is usually taken as that for which it is seen on 50 per cent of the presentations, this is the luminance L_T in Fig. 3.1. Fig. 3.1 is known as a 'cumulative frequency curve' or 'ogive'.

It is often stated that the eye has the ultimate sensitivity, in other words that it can detect single quanta. This is not precisely true. In fact a flash of light can just be detected if about ten quanta are absorbed over an area of the retina subtending $10'$ of arc within about $0\cdot1$ sec. It is estimated that this area contains about 500 rods, so the probability of one rod absorbing more than one quantum is very low. This means that a single quantum can activate a rod, but nevertheless a single stimulated rod by itself cannot produce a visual signal. It needs the almost simultaneous responses of more than one rod (perhaps as many as ten) to be added together somewhere in the visual system before the subject actually perceives the flash of light.

So far we have been discussing the absolute visual threshold, namely the minimum quantity of light required to produce a visual response in a given set of conditions. There are also visual thresholds related to the difference in appearance between two fields. For example if two adjacent fields are set at equal luminance they will appear as a single uniform

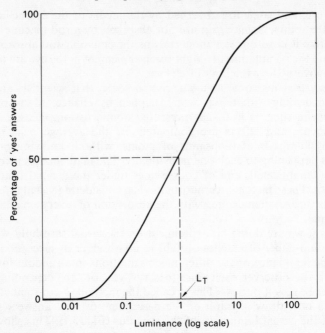

Fig. 3.1 Frequency distribution of a subject's response to the presence of a field near the threshold luminance.

field. If we now raise the luminance of one until we can just see a difference in luminosity this extra luminance is termed the luminance difference threshold. This is discussed further in section 3.7. Related thresholds for fields just differing in colour will also be discussed in section 10.4.1, p. 158. All such threshold measurements are subject to statistical fluctuations.

3.3 DARK ADAPTATION AND LARGE FIELD ABSOLUTE THRESHOLDS

Thresholds have been used extensively to study the sensitivity of the eye. A classical example is the dark adaptation function shown in Fig. 3.2 where changes in the absolute threshold are used to place a numerical description on the increased sensitivity of the eye as it adapts to darkness. This function is measured by first allowing the eye to reach an equilibrium state with a field of moderate to high luminance. This is known as light adaptation and may take about 5 min to complete. If the subject is next placed in the dark his eye will initially be relatively insensitive to light due to the bleaching of the receptor pigments. By measuring the absolute threshold luminance of a field the increase in sensitivity with time in the dark can be monitored. Sensitivity is often defined as the reciprocal of the threshold luminance (or energy).

Normally the dark adaptation function is found to be divided into two parts probably arising from the separate regeneration of rod and cone pigments. These can be seen in Fig. 3.2. The portions shown as broken lines are not measurable except under very special conditions. This is because the subject simply varies the threshold field until it is just seen. Since he has no way of distinguishing between a rod or a cone threshold the luminance is lowered until the most sensitive of these is just seen. Consequently only the part of the double function drawn as a continuous line in Fig. 3.2 is measured. It is a reasonable approximation to take the luminance threshold of the day retina (the cones) for a large field (i.e. 1° in extent or larger) as 10^{-3} cd m^{-2}, and that of the night retina (the rods) as 10^{-6} cd m^{-2}, although these figures depend upon the size of the stimulus and on the diameter of the eye pupil.

Strictly speaking this function only tells us how the sensitivity varies

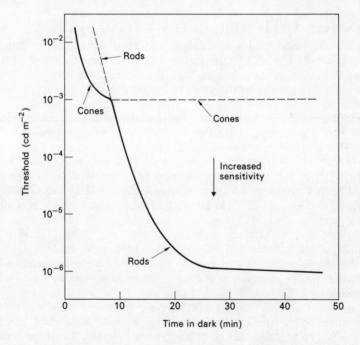

Fig. 3.2 The dark adaptation function (continuous line) of a light adapted subject. The broken lines are continuations of the separate hypothetical rod and cone functions.

at the visual threshold. However, it is at least qualitatively indicative of how the eye responds to higher luminances or supra-threshold stimuli. Hence, a light will have a higher luminosity when viewed by the dark adapted eye than when it is in a light adapted condition.

What is the visual process described by the dark adaptation function? It was originally thought that the threshold was a measure of the amount of visual pigment which remained bleached. Retinal densitometry has partly confirmed this view although it is the logarithm of the threshold which is proportional to the amount of bleached pigment. This technique has also shown that the cone pigment regenerates at a much faster rate than that in the rods. This could be deduced from Fig. 3.2 where it is seen that the cone function reaches equilibrium within approximately 7 min and the rod function within 30 min. Rushton found that for each 5 per cent of the rod pigment which is bleached by a light stimulus the threshold is raised by 1 log unit while for the cones the corresponding value is 0·15 log units. Part of the increase in visual sensitivity during dark adaptation is probably due to changes in the neural connections of the retina as the visual pigments regenerate.

3.4 SMALL FIELD THRESHOLDS—SEEING STARS

We are often interested in how faint a very small object can be when it is just detected. This is of importance when we want to know how far away a lighthouse can be seen at night, or how far away an aircraft navigational light can be detected, and of course how faint is the faintest star.

Since the threshold for point objects is dependent on the amount of light entering the eye, it is often quoted as the illumination of the eye pupil for a 50 per cent probability of detection of the source. This is called the *point brilliance* and the threshold value is about $3·4 \times 10^{-9}$ lux or $1·4 \times 10^{-11}$ W m^{-2}. With a normal pupil of diameter 8 mm under these conditions a power of 7×10^{-16} W enters the eye. The incredible sensitivity of the eye can perhaps be realized when it is stated that a 100 W lamp radiates about 1·5 W as light. In other words such a lamp emits sufficient light to produce a threshold light for 2000 million million people, which is about 600 000 times the number of people living today in the whole world. If a power of 7×10^{-16} W were supplied to heat 1 gram of water, it would take nearly 200 million years to raise its temperature by 1°C.

Looked at another way a threshold point brilliance of $3·4 \times 10^{-9}$ lux is the illumination the eye receives from a point source of luminous intensity 1 cd at a distance of 17 km (about 10 miles). In fact in order to achieve certainty of seeing, which is necessary when calculating the effectiveness of lighthouses and beacons, one works on a value of 100 times the threshold. This means a source of 1 cd can be seen with certainty at a distance of 1 mile.

The amazing sensitivity of the eye enables one to see very faint stars, although the number which we can see with the unaided eye on a clear night with no nearby glare is not as many as might be thought. In fact the number is of the order of 2000 which can be seen from a single location. The number visible in the whole sky under the most favourable conditions is about 6000. Star magnitudes are reckoned on a numerical scale, with increasing numbers for the fainter stars. Magnitude 1·0 corresponds to the mean intensity of the two nearly equally bright stars Altair and Aldebaran, whilst magnitude 6·0 stars are just on the visual threshold. The scale is a logarithmic one, the ratio of intensity for a difference of 1 magnitude being 2·512:1, and for a difference of 5 magnitudes, 100:1.

It is easy to carry out an experiment on one's rate of dark adaptation on a clear moonless night using a star atlas and a list of stellar magnitudes. Choose a site with no interference from street lights and equip a torch with a deep red filter. (This enables one to see the time and read the atlas and tables with minimum bleaching of the visual purple in the rods, since this is not very sensitive to light at the far red end of the spectrum.) It is only necessary at intervals of a minute or two to find a constellation at a fairly high altitude and to identify the faintest star seen. Typical results are shown in Fig. 3.3, and these were taken in southern England in September. (Note that this method is not sufficiently sensitive to determine the initial cone adaptation, see Fig. 3.2.) It is interesting that although it is often stated that one can see stars down to magnitude 6·0, it is not often made clear that this can only be achieved after having been in the dark for about 25 min. Fig. 3.3 also shows that on another night, a very slight haze reduced this threshold by about one magnitude.

Astronomers have shown that by sitting in a dark room having only a very small opening towards the night sky, stars of magnitude as low as 7 or 8 can be detected. This shows that the normal background luminance of the night sky is sufficient under these conditions to reduce the threshold in this way, although the practical difficulties of identifying a few stars seen through a narrow opening are very real.

The threshold is naturally very reduced if the luminance contrast is reduced between the object and its background. That is why bright stars easily visible at night quickly disappear at dawn and can never be seen by the eye alone during the day. Apart from the sun and moon only the planet Venus becomes bright enough at times to be seen in broad daylight with the unaided eye, providing one knows in which direction to look.

It is often stated that when the eye is dark adapted and the visual pigments fully regenerated the rods are much more sensitive than

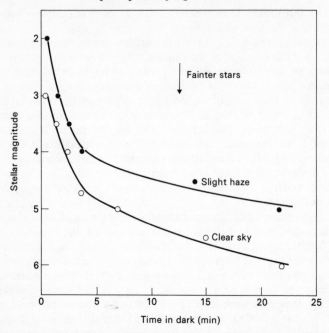

Fig. 3.3 Stellar magnitudes showing the faintest stars just seen by an initially light-adapted subject.

the cones. This is strictly true only for relatively large lights which stimulate large numbers of receptors. Individual receptor responses can be added together to produce a greater response so that with very small fields like stars which only stimulate a few receptors the threshold is much higher. This summation ability is more apparent with groups of rods than groups of cones, and with point or star-like stimuli the rod threshold is only a little lower than the cones. Thus, the main contribution to the increased sensitivity of the rods, as the eye becomes dark adapted, as can be seen for a large light in Fig. 3.2, is probably due to area summation perhaps arising at the horizontal cell level in the retina connections (Fig. 2.2) and not from differences in response of individual rod and cone receptors.

A small astronomical telescope will reduce the threshold considerably. It is easy to show that a telescope will not increase the apparent brightness of an extended object such as the moon. In fact it appears slightly dimmer through the telescope; this can easily be verified by looking at the moon through a telescope with one eye and simultaneously viewing it with the other unaided eye. The reason is that the moon appears magnified through the telescope, and the extra light collected is spread over a larger image. The two effects compensate, except for the slight dimming which is due to light lost in the lenses and by reflection at their surfaces. The situation is, however, very different when viewing a star. This is so far away that it will appear to be a point no matter what the magnification, providing it is not so great that the spurious disc image due to diffraction becomes apparent. Provided then that the magnifying power is sufficiently great for all the emerging light from the eyepiece to enter the eye, the luminosity of a star is increased in a manner dependent on the ratio of the area of the objective lens of the telescope to the area of the eye pupil. Since the latter is about 50 mm^2 at night, the increase in the retinal illuminance is (area of objective in mm^2)/50. Thus for a 75 mm diameter objective, the increase is 88 times. This corresponds to a magnitude difference of 4·9, enabling stars of over magnitude 10 to be seen, this is allowing for some absorption of light in the telescope.

The brighter stars and the planets (including the elusive Mercury) can be seen during the day with a small equatorially mounted telescope (60 or 70 mm diameter) which can be set in the correct direction with the aid of divided circles, providing the sky is not hazy, and the object is not too near the sun. This is because the star or planet is a small object and appears much brighter through the telescope, whereas the background sky is an extended object, and thus appears slightly darker. The contrast is thus greatly increased.

3.5 SENSITIVITY TO COLOURED LIGHT

The human eye is sensitive to only a very narrow band of the electromagnetic spectrum, in fact just about an octave (frequency ratio of 2:1) from approximate wavelengths of 380 nm in the violet to 760 nm in the red. (Compare the ear with a frequency range of about 7 octaves.) Fig. 3.4 shows how the relative thresholds for extended sources vary with wavelength, both for day or photopic vision (using cones), and for night or scotopic vision (using rods).

The sensitivity of the two receptors at different wavelengths can be measured by the reciprocal of the two threshold functions as shown in Fig. 3.4. These are known as the relative luminous efficiency (V_λ) functions for photopic and scotopic vision. Often the functions are normalized so that V^λ is equated with unity at the most sensitive wavelength (555 nm for cones and 505 nm for rods), but such a graph is not so useful. For example it would not show—as can be seen directly in Fig. 3.4—that for large fields the rods are more sensitive than the cones except at the longer wavelengths. Very similar functions can be measured at suprathreshold levels if instead of measuring the energy to produce a threshold at each wavelength we measure the energy to produce a

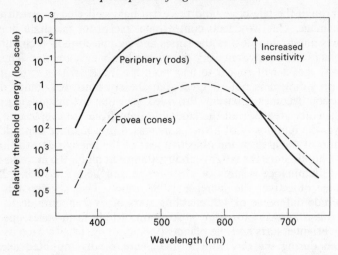

Fig. 3.4 Relative luminous efficiency functions for rod and cone vision measured at the absolute threshold.

given level of luminosity. At a high luminosity level only the cone function would be found and at low levels only the rod function.

Consequently in the daytime the eye is most sensitive to apple-green light of wavelength 555 nm, whereas at night the peak moves over to the blue-green at a wavelength of 505 nm. This shift of colour of the peak sensitivity is called the Purkinje effect, after the discoverer, who in 1825 noticed that the surfaces of signposts which were painted red and blue and which had the same luminosity in the daytime, looked different at dawn. The blue then appeared brighter than the red.

It is interesting that in sunlight the most energy is given at the wavelength at which the eye is most sensitive, which suggests that evolution has played a part. This evolutionary element is borne out by the fact that deep sea fishes have a peak sensitivity moved over to the blue, since sunlight filtered by a great depth of the ocean is blue.

3.6 PHYSIOLOGICAL BASIS OF LUMINOSITY

Before we discuss further aspects of luminosity it is helpful to take a look at the physiological evidence. Many studies have been done by inserting electrodes into simple eyes such as that of Limulus, the Horse Shoe Crab, and into the retinae of cats, monkeys and many other species. Although it is fairly well established that the actual receptors produce steady electrical potentials, in some way related to the light intensity falling on them, the actual signals sent to the brain are coded in a binary manner. A series of electrical action potentials or spikes are sent from the ganglion cells along their axons, the optic nerves, whenever the retina is stimulated. As the light falling on the receptors is increased, more of these spikes are produced in the same time. In other words the spike frequency increases, and can be as high as 120/s (Fig. 3.5). Luminosity is therefore frequency coded and the brain thus interprets a higher spike frequency as a higher intensity light. Furthermore the spike frequency is related roughly to the logarithm of the retinal illuminance. As was explained in section 2.3.2, p. 15, these action potentials are transmitted along the nerve fibres without decrement, which is obviously a great advantage for an information transmitting channel. Furthermore, as we know from frequency modulated radio transmission, the reception suffers from far less interference than does an amplitude modulated transmission.

The situation is, however, not as simple as this and although some

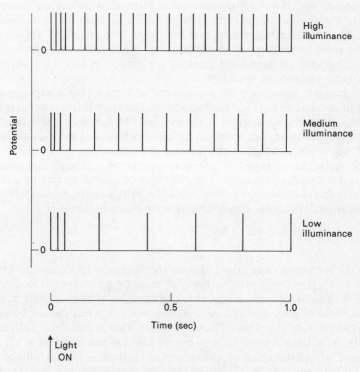

Fig. 3.5 Action potential responses measured from the eye of the Limulus at three levels of illuminance.

ganglion cells fire when the light is turned on, some only become active when the light is turned off. It is interesting to digress just for a moment to consider other senses, for example hearing, in which the sensation of loudness is similarly coded and sent to the brain as action potentials in no way different from those of the eye. It is the different location at which they finally arrive in the brain which results in the entirely different perception.

3.7 LUMINANCE DIFFERENCE THRESHOLDS

As we said earlier seeing detail and transmitting information is mediated by differences of contrast, and that the eye has an in-built mechanism for increasing apparent contrasts. We are therefore interested in finding out the luminance difference thresholds which the eye can perceive in various circumstances. This will enable us to find out whether an object can or cannot be seen against its background.

It will also enable us to discover how far away an object can be seen against the sky. If a black object recedes from us its luminance increases due to the air-light or light scattered into our eyes from the air molecules and dust particles between us and the object. Eventually it will appear of the same luminosity as the horizon sky. When it is at such a distance that it appears to have the threshold contrast with the sky, that is its maximum distance of visibility.

To measure luminance difference thresholds we set up a large circular field of luminance L, (Fig. 3.6a) and in the centre of this is a circular test patch subtending an angle θ at the eye, also of luminance L. Since this patch has the same luminance as its surround we will see a single large field. However, if we gradually increase the centre luminance to $L+\Delta L$, where ΔL is the increment in luminance, the patch will be distinguishable when ΔL is sufficiently large. The difference of luminance can be related to the general luminance (often called the light adaptation level) by the following relationship, which defines the luminance contrast C

$$C = \frac{(L+\Delta L)-L}{L} = \frac{\Delta L}{L}$$

and the threshold value of ΔL defines the threshold luminance contrast.

In fact it is not very precise simply to raise the central patch luminance until it can just be seen. In fact it is slightly more precise to start from a large value of ΔL and reduce it so that the patch just cannot be seen, and then to take the mean of the two values. This is called the method of limits. A better but more lengthy psychophysical procedure is to use the method of constant stimuli (as explained in section 3.2). A number of stimuli (perhaps 10 or 12) all of different luminance are presented in succession to the observer. Some are above and some below the threshold

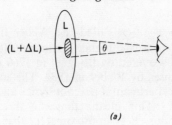

(a)

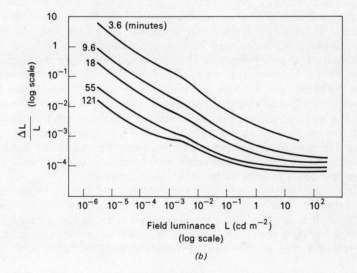

(b)

Fig. 3.6 Weber's fraction $\Delta L/L$ for different field sizes. Weber's law is seen to hold only at high field luminances.

value, and they are presented in random order. The observer reports either 'yes' or 'no' depending upon whether he sees it or not. A frequency ogive is then plotted and the 50 per cent probability of seeing point is noted. Blackwell produced some very comprehensive data for central fields of different sizes using binocular vision and a 10° diameter adapting surround field. His results are shown in Fig. 3.6(*b*). It will be seen that under the best conditions C is 0·008 or 0·8 per cent. This occurs with a 2° diameter field and a surround luminance of about 100 cd m^{-2}. For absolute certainty of seeing, these values of C can be multiplied by a factor of about 100. The discontinuity in the curves at about 10^{-3} cd m^{-2} marks the transition from rod to cone vision.

3.8 WEBER'S LAW

The value of $\Delta L/L$ is often called the Weber fraction. That this is approximately constant at the higher luminances (the nearly horizontal parts of 3.6(*b*) was discovered by Bouguer in 1760 using candles. It was extended to other senses by Weber in 1834. The so-called Weber law is one of the oldest in experimental psychophysics and in effect stated that $\Delta L/L$ is a constant. This means that the threshold is a constant fraction of the stimulus. It is approximately true at high values, and pervades the whole of human experience. One candle introduced in a dark room makes a great deal of difference, but is not noticed when introduced into a brightly lighted room. Thus the range of domestic electric filament lamps which are available (25, 40, 60, 100, 150 W) are spaced in a fairly large steps with an approximate ratio of $1:1\cdot5$.

Visual photometry depends upon the good luminance discriminating power of the human eye. It is a good method but is more time-consuming, and the measurements have less precision (i.e. more spread) than those derived from photo-electric instruments. The latter, however, are in fact used to imitate the eye and have to be calibrated for spectral sensitivity in terms of visual measurements (Fig. 3.4). They cannot be more *accurate* than the eye. In this sense it is wrong therefore to quote luminance measurements from photo-electric instruments to more decimal places than one could achieve by using a visual photometer.

It is now thought that visual discrimination is limited by the 'noise' of the retina. This is the random firing of visual cells even in the absence of light. It is easy to detect the perceptual effect of this by sitting in a completely dark room for 30 min or so. As time goes on one is increasingly aware of a scintillating greyness, the so-called 'dark light' of the retina.

3.9 FACTORS AFFECTING LUMINOSITY

In assessing any visual situation we are initially interested in how bright surfaces appear to us. Many factors can cause a variation in the so-called luminosity and we will now consider some of these.

Luminosity is sometimes defined as the subjective correlate of luminance or retinal illuminance. This is a useful definition in the sense that if all other factors remain constant the luminosity and luminance increase together. Fig. 3.7 shows a hypothetical function of this type which we can call a scale of luminosity. At low luminances below the visual threshold, the luminosity of the field is constant corresponding to the dark light of the eye, above threshold this increases together with the luminance. What is of greater interest, however, is how the luminosity varies when factors other than the field luminance are also allowed to

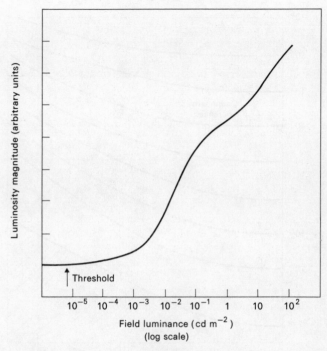

Fig. 3.7 A hypothetical luminosity function.

vary. For extended fields the most important factors are the luminance to which the subject was preadapted, the luminance of any surround field and the colour of the field.

3.9.1 Preadaptation

The major temporal effects of previous light adaptation are shown by the dark adaptation function discussed in section 3.3. As we have seen, light adaptation lowers the sensitivity of a test field viewed in the dark for several minutes after removing the preadaptation stimulus. This is indicative that at suprathreshold levels the luminosity is similarly lowered. Fig. 3.8 shows the influence of a range of preadaptation luminances on the luminosity of a foveal test field, at and above threshold level, immediately after the preadaptation luminance is removed. Each contour was obtained for one luminosity level and shows by how much the test field has to be varied to maintain the luminosity constant as the preadaptation luminance is varied. As the preadaptation increases in general the eye becomes less sensitive and the test luminance must be raised. The higher contours refer to higher levels of luminosity.

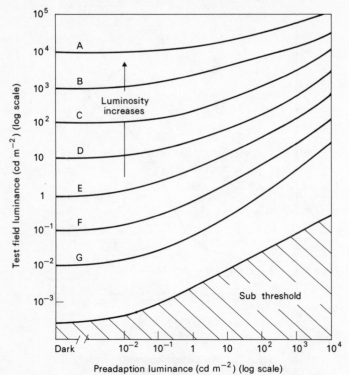

Fig. 3.8 Contours of constant luminosity measured with a dark background with the subject preadapted to a range of luminance levels.

3.9.2 Induction and surround fields

We have already seen in section 3.1 that an adjacent or a surround field can influence the appearance of a test field. The effects at different luminance levels are indicated by the equal luminosity contours of Fig. 3.9, obtained for foveal vision. These are similar to contours of the previous Fig. 3.8 although closer inspection shows that the influence of a surround field is greater than a preadaptation field of the same luminance. Consider for example the conditions determining contour *D* on each figure. These both refer to the same luminosity level which is approximately equivalent to the luminosity of a piece of white paper viewed under any common daytime illuminance. For dark preadaptation or for a dark surround the adaptation conditions are identical. If the subject is now preadapted to a luminance of 100 cd m^{-2} the test field must be raised to 180 cd m^{-2} to produce the same luminosity. However, if the surround is raised to 100 cd m^{-2} the test field must be increased to a much higher luminance, namely 550° cd m^{-2} to produce the same luminosity.

Frequently the preadaptation and surround luminances are similar but what happens if these differ considerably? One might, for example, look through a window, from a highly illuminated sunlit scene to an object in a dimly-lighted room. To a useful approximation it is the surround luminance which moderates the luminosity unless this is considerably less than the preadaptation level (perhaps 100 times smaller). A second consideration is what happens when the luminous distribution of the surround varies from area to area. For example the reader may be seated at a desk with many other objects in the total field of view all of which contribute to a complex surround. In many cases the variation in luminance may not be as complex as it first seems. The reflectances of objects probably do not vary by more than 20:1 and if the illuminance is fairly even the luminances will vary by a similar amount.

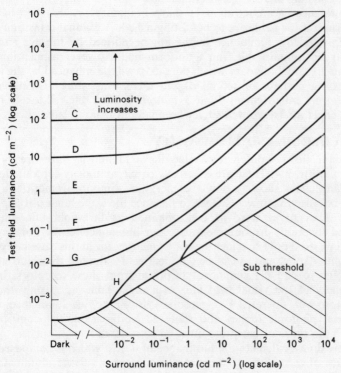

Fig. 3.9 Contours of constant luminosity measured for a range of background luminances.

In this case the data of Fig. 3.9 will adequately define the luminosity if the surround luminance is equated with the mean luminances of the surrounding objects. However, when the luminances in the surround vary by large amounts this simple if approximate approach is not viable. A common example is the effect of an oncoming car headlamp when driving at night. Here we have a field of high luminance (sometimes called a glare source or an induction field) in an otherwise near dark background. Any other object in the scene such as a signpost will have its luminosity changed (normally lowered) by the presence of this source. This is partly the inhibitory effect of the retinal receptor pathway responding to the glare source on the receptor pathway responding to the test object, and partly due to light from the glare source being scattered in the eye, then being absorbed by all parts of the retina and hence lowering the sensitivity of all receptor pathways. In general this inhibitory effect is greater for glare sources of larger size and of higher luminance which are closer to the test object in the field of view.

3.9.3 Contours and spatial discrimination

The presence of a contour or a sharp boundary between two fields is important if the fields are to be distinguished. A gradual transition from a high to a low luminance might not be noticed by the eye and can appear indistinguishable from a uniform field. However, as mentioned in section 3.1, any boundary which does exist will be enhanced by the visual processing perhaps of retinal origin, thereby assisting discrimination. This border enhancement (Plate 3, *bottom*) is often referred to as a Mach band effect after the discoverer who analyzed many related effects.

3.10 COLOUR AND LUMINOSITY

The main effects of different wavelengths on luminosity have already been shown in Fig. 3.4. These are effectively equal luminosity contours for the rods and cones showing how the energy of a monochromatic test field must be varied with wavelength to produce the same luminosity appearance. The curves were in fact obtained at threshold but similarly shaped functions are obtained at higher luminance levels. For daytime vision we need only consider the cone function, for in this case the eye will be light adapted and as can be seen in Fig. 3.2 the rod threshold will be too high to cause a significant response from these receptors. In the periphery of the visual field and with night time vision only the rod function will operate. For mesopic vision, i.e. at luminance levels between photopic (cone) and scotopic (rod) vision, the contribution from both types of receptors must be considered.

How do we calculate the luminance of a field which is composed of a mixture of a number of wavelengths such as a surface illuminated by daylight or a tungsten lamp? In this case we simply add the separate contributions of each wavelength (or narrow band of wavelengths) to obtain the total luminances and assume that luminance has an additive effect on luminosity. Hence the luminance of an object is defined by the equation:

$$L = k \sum E_\lambda V_\lambda$$

where k is a constant and E_λ is the energy emitted in each band of wavelengths λ, which have a relative luminous efficiency of V_λ. Luminance is therefore additive by definition and the assumption that equal luminances produce similar luminosities irrespective of the wavelengths present is a good approximation to practice. In general, however, the more a test field approximates to a monochromatic field the more the luminosity increases. This phenomenon where a white field (which must contain a mixture of wavelengths) appears to have a lower luminosity than a monochromatic field at the same luminance level is known as the Helmholtz–Kohlrausch effect (see also section 8.4, p. 135).

3.11 MEASUREMENT OF LUMINOSITY

Since luminosity is an attribute of perception which has magnitude it can be scaled as shown in Fig. 3.7. For example we can place a series of sources or surfaces in order from the dimmest to the highest luminosity. However, whether the corresponding magnitudes can be given numbers such that, for example, one field can be said to be twice the luminosity of a second field has remained a controversial question of psychophysics for more than a century.

Fechner, who has been described as the father of psychophysics, tried to overcome this problem by assuming that the perception of the luminance difference threshold between two fields (often called a just noticeable difference or jnd) is a unit of luminosity. Hence if we increase the luminance of a field by a certain amount and count the number of jnds, we have a measure of the change in the magnitude of the luminosity. Assuming that Weber's law is true (section 3.8) Fechner was able to show that the luminosity increases with the logarithm of the luminance. In fact, as can be seen from Fig. 3.6(b), Weber's law does not hold at low levels of luminance so that a logarithmic law can only be expected at the high luminance levels. Fechner's luminosity function is shown in Fig. 3.10 using the data for the lowest curve of Fig. 3.6(b). A stepped function has been drawn to indicate that luminosity increases in steps as each successive jnd is reached. The function follows a logarithmic trend in the region where Weber's law is approximately true.

Fechner's logarithmic law has never been fully accepted because the size of a jnd of luminance varies from subject to subject and also depends on the colour, size and shape of the field. One way around this difficulty

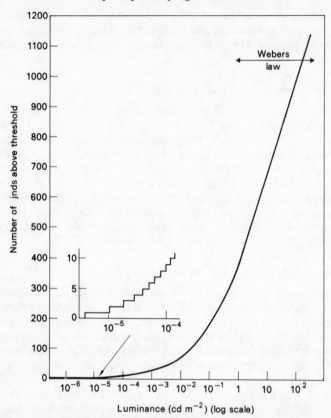

Fig. 3.10 Fechner's luminosity function obtained from the data of the lower curve of Fig. 3.6. Fechner's logarithmic law which predicts a straight line function is seen to hold where Weber's law is correct. The insert indicates the initial hypothetical rise in luminosity as the luminance is raised by each jnd.

would be to change the size of the luminosity unit when dealing with different subjects, or with fields of different size, shape or colour. However, many visual scientists have found this an unjustifiable procedure, although it might be argued that we do not refuse to use the kilogram simply because a unit mass of lead occupies a different volume from the same mass of water. Fechner has also been criticized for his integration of jnds to obtain the logarithmic law but this can be overcome by using the law to show the trend of the luminosity function rather than its detailed shape. Finally there is no answer to the criticism that the jnd may not produce a unit change of perception.

A second method of scaling luminosity is the partition method. This method was first used by Plateau a century ago. He asked a number of artists each to paint a grey surface which appeared to be half way between black and white. To his surprise each artist chose a grey of about the same luminosity. A related technique was used to obtain the Munsell scale of Value which will be discussed in more detail in section 6·33, p. 96. A subject is simply asked to choose a set of luminances which appear to vary in equal steps of luminosity, the steps being much greater than 1 jnd of luminance change. For example the subject might be presented with a series of grey cards varying in luminosity from black to white. Ten cards are then chosen from perhaps 50 or more which when placed in order increase in equal steps of luminosity. These chosen cards are given the numbers 1 to 10. The difference between two numbers is then a measure of the difference in luminosity of the corresponding cards although number 4, for example, is not twice the luminosity of card number 2. What is particularly interesting is that the number of jnds between successive cards has been shown to be constant as expected if Fechner's assumption is correct.

A third method of measuring luminosity which has become increasingly popular in recent years, mainly due to Stevens at Harvard University, is the method of direct magnitude estimation. The method however is not new and has been applied to light sources for nearly 2000 years, since the second century B.C., when the Greek astronomer Hipparchus divided stars visible to the naked eye into six classes or magnitudes, according to their luminosity, the brightest being of the first magnitude and the faintest comfortably visible on a moonless night, of the sixth magnitude. Fortunately his estimates were preserved by Ptolemy, who in his Almagest gave the positions and magnitudes of some 1000 stars. The magnitude estimates of Hipparchus can be fitted to an approximately logarithmic scale based on accurate modern instrumental observations. Thus, equal ratios of intensities are represented by equal magnitude differences. Stevens has suggested that an observer can go further than simply classifying luminosities into a number of groups, and that direct estimates of the magnitude of a luminosity can be made.

These direct estimates can be made by anyone using a very simply constructed apparatus shown in Fig. 3.11. Neutral filters can be used to regulate the luminance of the stimulus. They can be obtained using cinemoid Pale Grey No. 60* which has an optical density† of approximately 0·5. Densities are additive so that two sheets superimposed give a combined density of 1·0, and three 1·5, and so on. An observer is presented with a number of luminances at random, and asked to give a number to represent its luminosity. He is told that it does not matter which number he gives to the first, but that subsequent numbers he gives should be related, so that if the second appears half the luminosity of

* See Appendix p. 183.   † See Glossary p. 180.

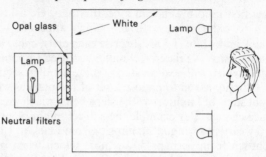

Fig. 3.11 A simple apparatus for making magnitude estimations of luminosity.

the first, the second estimate given should be half that of the first. It is best to replicate each luminance 5 or 10 times and then to take the mean estimates for each luminance. In order to normalize the readings from different observers, it is usual to divide all readings by the mean estimate of a standard stimulus (e.g. 500 cd m^{-2}) seen in the dark. This has been done for 200 observers and the results plotted in Fig. 3.12. The results are surprisingly consistent, although the precision is not comparable with that attained by simultaneous photometric matching. Under conditions of dark adaptation the results can be represented approximately by a power law of the type:

$$M = S^n,$$

where M is the magnitude estimate, S the stimulus magnitude, and n is an exponent quoted by Stevens as 0·33, but by the authors as 0·26. Sometimes the results are better represented by:

$$M = k(S - S_o)^n,$$

where k is a constant and S_o is the effective absolute threshold value of the stimulus.

This type of function has not gained general acceptance because, like Fechner's logarithmic law, it is subject to considerable variations between subjects.

What is the reader to make of this? Do any of the three methods discussed above allow us to scale luminosity or not? Do any of the numbers add meaning to the luminosities other than to allow us to place them in the correct order? Perhaps not, but nevertheless the methods can be useful for recording perceptual changes especially when comparing luminosities under different visual conditions, for example, under different adaptation levels, as shown in Fig. 3.12. Attempts have been made to test the results of these methods by comparing the luminosity scale with the electrical activity in the eye and brain. Sometimes poten-

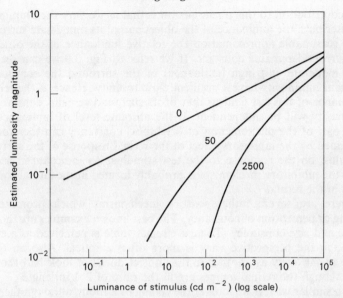

Fig. 3.12 Luminosity functions obtained by magnitude estimations at three background luminances of 0, 50, and 2500 cd m^{-2}.

tials have been found which increase with the logarithm of the luminance (or retinal illuminances) and sometimes in the form of a power function. In any case these responses are probably occurring at an early stage of the visual analysis and further modifications will occur before the final perception of the luminosity results.

3.12 BRIGHTNESS (OR LUMINOSITY) CONSTANCY

The eye works in such a way that as we approach or recede from a surface, its luminance remains constant. This is because when we get near to an object, although more light reflected from it enters the eye, it is spread over a larger retinal image. This has the advantage that as we move about in our environment, the luminances and therefore the luminosities remain constant.

Another example of the constancy of luminosity which is normally called 'brightness constancy' is that a piece of white paper remains white whether viewed in sunlight or in a moderately illuminated room. Similarly a piece of coal looks black under both these conditions even though the luminance of the coal in sunlight might be 100 times the luminance of the paper in the room. If luminance determines the magnitude of luminosity we would expect these appearances to be reversed.

One contribution to this phenomenon is that as we vary the illuminance we alter both the luminance of the object and of its immediate surround. To a reasonable approximation the relative luminance of the object to the surround remains constant. If we refer to Fig. 3.9 we can see that with moderate and high luminances of the surround the contours of constant luminosity have a gradient close to unity. Hence, if the relative luminance of the test field to that of its surround remains constant the luminosity will be independent of the absolute level of luminance. At least part of the phenomenon of brightness constancy can therefore be explained by the inhibitory effect of the retinal response of the surround stimulus, on the response to the test stimulus. As indicated in section 3.9.2 the inhibitory mechanism is probably located in the lateral connections in the retina.

There are several other sensory mechanisms which promote the feeling or perception of constancy. The best known examples are those of shape and size constancy. Thus, a circular table is perceived as a circle, although the perspective view is more often elliptical. Also an object twice as far away as another similar object does not look half the size, even though the retinal images are in the ratio of 2:1 in length.

In a similar way, if we look at the ceiling of a room, although the parts near a window look of higher luminosity than those in the far corners, we actually 'see' it as much more constant than would be measured by a photometer. This has sometimes been called 'seeing through brightness to reflection factor' and is another example of luminosity constancy. This obviously helps us to see a uniformly decorated wall or ceiling as an entity, and to minimize the effect of local variations of illumination. These mental stabilizing influences all tend to reduce the otherwise almost overwhelming, rapidly varying amount of information which the human receives all the time from his environment. The perception of a reasonably stable environment helps in the assessment of a given situation. Otherwise it would be too confusing to live for more than a few moments.

4
What is colour?

4.1 NEWTON'S EXPERIMENTS

The nature of colour was first discovered by Sir Isaac Newton (1642–1727), and although this fact is well known, the details of his work are not, and are certainly very instructive to repeat. Unlike earlier philosophers, and genius though he was, he was very ready actually to carry out experiments, and if necessary to make things with his own hands.

He resided in Trinity College, Cambridge, from 1661 to 1701. His active interest in optics began in about 1663 when he started grinding lenses and working on the construction and performance of telescopes. This led him in 1666 to buy some prisms at Stourbridge Fair, near Cambridge, to 'try therewith the celebrated phenomena of colours'. Obviously the colours produced by passing light through a prism were known at that time but the proferred explanations were quite incorrect. Apparently Newton made no discoveries at first, but in 1672 he sent a scientific paper to the Royal Society entitled 'Theory of Colours', in which he gave an account of his prism experiments.

He continued the experiments in optics and they were completed and almost written up in book form in 1692. However the well-known and often embellished incident occurred when all his notes and manuscripts were destroyed by a fire. Whether his subsequent illness was a direct result is not clear, but the fact is that he did not re-write and publish the book until 1704 when it appeared as the monumental *Opticks*.

He started his experiments on colour by placing a prism in front of a hole in a shutter of a darkened room. A beam of sunlight entered the hole, passed through the prism and was allowed to fall on a sheet of white paper on which it appeared as a coloured spectrum band (Fig. 4.1). In his own words:

> The Spectrum did appear tinged with this Series of Colours, violet, indigo, blue, green, yellow, orange, red, together with all their intermediate Degrees in a continual Succession perpetually varying so that there appeared as many Degrees of Colours as there were sorts of Rays differing in Refrangibility.

Newton was convinced that these colours were already present in the original white light, and were not introduced by the prism as was then

61

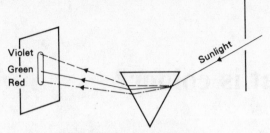

Fig. 4.1 Newton's experiment.

thought. To check this he first recombined the coloured lights formed by the prism, and this he did by two quite different methods: first using a lens (Fig. 4.2) and secondly using two further prisms (Fig. 4.3). The result in both cases was white light similar in every way to the original. He then placed a second prism immediately after the first in the original experiment, but at right angles to it, so that the light passed through both prisms (Fig. 4.4). If the first prism was used alone it would form a vertical spectrum S_1 on the screen, whereas the second prism would form a horizontal spectrum S_2. When used together, Newton saw that a spectrum was formed along S_3, inclined to the vertical. This showed

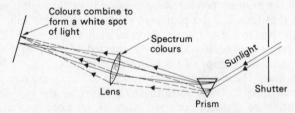

Fig. 4.2 The recombination of colours using a lens.

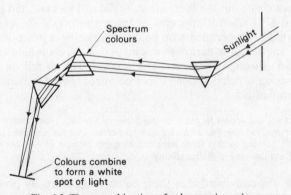

Fig. 4.3 The recombination of colours using prisms.

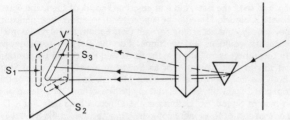

Fig. 4.4 Newton's further experiment.

quite conclusively that white light could be split up into lights of different colour by a prism, but that these different coloured lights could not themselves be split further. If this were possible, the violet light V in the spectrum S_1 formed by the first prism would be spread out into still further colours along the straight line VV^1 when refracted by the second prism. Instead the violet light is refracted unchanged to V^1.

Newton thus realized that white light was a composite mixture of different kinds of rays. We now know, of course, that they are electro-magnetic waves which differ in frequency. The eye is sensitive to light of different frequencies and perceives these as different colours. Table 4.1 gives the frequencies and wavelengths of light and their corresponding colour perceptions.

Newton realized that colour is a perception and needs an individual to

TABLE 4.1

Colour	Wavelength nm	Frequency Hz × 10¹⁴
Violet {	400	7·5
	450	6·7
Blue	480	6·2
Blue-Green	500	6·0
Green	540	5·6
Yellow-Green	570	5·3
Yellow	600	5·0
Orange	630	4·8
Red	750	4·0

receive the rays of light and to interpret them as colour. The actual rays themselves are no more coloured than are radio waves or X-rays, but they have the necessary message or information to enable colour sensa-tions to be evoked. This is clear from the following extract from the *Opticks*:

The homogeneous Light and Rays (i.e. monochromatic light) which appear red, or rather make Objects appear so, I call Rebrifick or Red-making; those which make Objects appear yellow, green, blue and violet, I call Yellow-making, Green-making,

Blue-making, and so of the rest. And if at any time I speak of Light and Rays as coloured or endued with Colours, I would be understood to speak not philosophically and properly, but grossly, and according to such Conceptions as vulgar People in seeing all these Experiments would be apt to frame. For the Rays to speak properly are not coloured. In them there is nothing else than a certain Power and Disposition to stir up a Sensation of this or that Colour. For as Sound in a Bell or musical String, or other sounding Body, is nothing but a trembling Motion, and in the Air nothing but that Motion propagated from the Object, and in the Sensorium 'tis a Sense of that Motion under the Form of Sound; so Colours in the Object are nothing but a Disposition to reflect this or that sort of Rays more copiously than the rest, in the Rays they are nothing but their Dispositions to propagate this or that Motion into the Sensorium, and in the Sensorium therefore Sensations of those Motions under the Forms of Colours.

4.2 LIGHT SOURCES AND THE PRODUCTION OF COLOURS BY SELECTIVE REFLECTION AND ABSORPTION

The all-important conclusion from Newton's work was that the colours were present in the original white light, and that the prism was instrumental not in producing the colours, but only in separating them out. The perception of colour can be split up into a chain of events. First, we start with a lamp or source emitting light, let us say white light (i.e. all colours mixed together). Normally of course in everyday life we do not see colours produced by prisms but by the simple action of coloured surfaces and glasses. These contain dye or pigment which act differently for different spectral colours. For instance if white light falls on a 'red' surface (Fig. 4.5(a)) this reflects the red light strongly but absorbs the remainder of the spectrum. This process is called selective reflection.

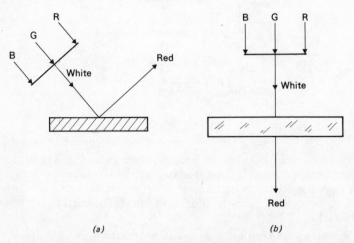

(a)   (b)

Fig. 4.5 (a) Selective reflection. (b) Selective absorption. For simplicity white light is shown only as red, green and blue, although all colours are present.

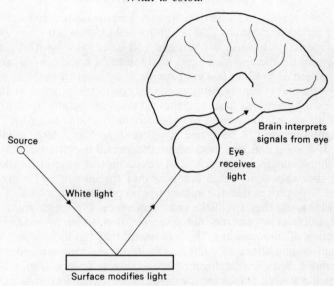

Brain interprets
signals from eye

Eye
receives
light

Source

White light

Surface modifies light

Fig. 4.6 The colour perception chain.

Similarly a red filter (Fig. 4.5(*b*)) absorbs all the colours of the spectrum except red, which it transmits. This is called selective absorption. We can now see the whole chain of events in the perception of colour (Fig. 4.6). The emission of light and its modification by selective reflection and its progress to the eye are purely physical phenomena, and up to this point there is no colour. As Newton said 'the Rays are not coloured'. But when this modified light enters the eye it is received by the retinal cones which absorb some and in doing so give rise to nerve signals which are interpreted by the brain as a colour perception. This involves a very complex mechanism which will be discussed later in this chapter. It is, however, opportune to consider why we have colour vision, and it is obviously necessary at this point to consider human vision in the context of vision generally in the animal kingdom.

4.3 ANIMAL COLOUR VISION

It is not, of course, an easy matter to establish whether an animal sees objects in colour, and although many lengthy and patient experiments have been carried out on many animals, there is still a great deal unknown. The methods are based on behavioural training, and often consist of allowing the animal to enter a box in one side of which there are two differently coloured papers or lights in panels, where there are also equal rewards of tasty morsels of food. This balances the olfactory cues. However, if say, when the animal goes towards a red light he is given a mild electric shock and he cannot actually reach the food, he can be trained to go towards a blue light and be rewarded with food. The positions of the red and blue lights with their shock and rewards are then interchanged at random to avoid simple directional learning. It may be that the animal is simply being trained to go to the brighter of the two lights as it appears to him, and that he may not be able to detect any colour difference. In order to see whether this is so, the luminance of the two lights then has to be varied stage by stage over a large number of trials. If a point can be found where the animal is confused, that is his equal luminance point, which is of course not in general the point at which they appear of equal luminosity to the human or to any other animal. This proves that the animal has no powers of distinguishing red from blue when they are of the same luminance. If no such point can be found, he obviously has red-blue discrimination, which proves he has the beginnings of colour vision. The experiment then has to be repeated with different combinations of colours. Obviously the tests are very lengthy and tedious. With less intelligent animals such as frogs a drum is rotated on which coloured stripes are displayed. Their reactions can be noted by observing eye or head movements.

The facts as we know them are briefly as follows. The amphibia have poor or no colour vision, and since they are mostly nocturnal and secretive, it is perhaps not surprising, since colour vision would be of little use to them. Of the reptiles, snakes probably have no colour vision, whereas it is thought that lizards have some, and turtles certainly have it to some degree. Strong diurnality is uncommon in mammals, and most have no colour vision, except for the primates. Cats probably do not have colour vision, whereas dogs have a weak colour sense. Cattle have no colour vision and a red rag is not red to a bull. All fish have either very good or some kind of colour vision, as do the diurnal birds. The lower primates are mainly nocturnal without colour vision. When we come to the higher primates, however, it is quite a different story, and the following have a colour vision mechanism which is almost identical to that of the human. They include the chimpanzee, Guinea baboon, pig-tailed macaque monkey, rhesus monkey, sooty mangabey, squirrel monkey, and the spider monkey. It appears that good colour vision goes with a well-developed and efficient eye which means a good accommodation mechanism for accurately focusing for clear vision. In other words it is developed in those animals whose activities, environments, and brains need and can take advantage of very good vision. Vision capable also of colour discrimination enables the animal not only to distinguish objects which differ in luminosity (which can be done with a simple black-white mechanism) but also enables distinction to be made between objects of the same luminosity which differ only in colour. A simple and obvious example is the power of being able to distinguish between ripe and unripe fruit (such as between red and green tomatoes), whereas this would not be possible with monochrome vision. What this means is that the animal with colour vision can receive more information about his environment.

It is easy to understand why birds need to have the maximum information input because they move so quickly through their environment when in flight, and they also need to see their food and prey from a considerable height.

Insects and birds wage war by camouflage, and a caterpillar disguised as a leaf obviously has a better chance of survival than one without the disguise.

Fish presumably need colour vision because luminosity contrasts are low in their murky environment, and colour contrasts then assume greater importance than in the clearer visual surrounds of dry land. Undoubtedly too for some species, the recognition of coloured spots on the mate plays an important part in breeding.

When we come to the higher primates and man, it is of course their possession of larger and more complex brains which enable them to take advantage of the increased information available from their colour-conscious eyes. The fact that in the human colour assumes a great aesthetic importance is of course a bonus, but probably purely an accidental one. Colour to most people plays a very essential and often very pleasurable part in their lives, and man being inquisitive has devoted much time speculating on its origin and meaning. Wonderful though it appears, the detailed facts are even more intriguing. Before we consider the theories of colour vision it is necessary to explain in more detail some properties of human colour perception.

4.4 FURTHER FACTS ON HUMAN COLOUR VISION

The magic number in the case of normal human colour vision is three, a fact which was first discovered by the versatile genius Thomas Young (1773–1829). He earned his living as a practising doctor in London, and never taught at University. All students of science have heard of Young's modulus, Young's slits, and his wave theory of light, but it is perhaps not so widely known that he was also famous for his many achievements in Egyptology, amongst which was his contribution to the cracking of the code to the Egyptian hieroglyphics.

His name still features prominently whenever colour vision is discussed even to this day, and the irony of his genius is that his main ambition in life was to achieve recognition and fame as a physician, not as a scientist. His famous colour mixing experiment is one of the most beautiful in the whole of science, and should be repeated by everyone. Three slide projectors are needed, and red green and blue filters (see Appendix for experimental details). These are arranged as shown on Plate 4 so that overlapping circles of light are projected onto a screen. An overlap means addition, so where the red and blue overlap we sum up in the eye the effects of red and blue light, and see a magenta colour. Similarly the blue and green overlap gives a cyan colour. These two mixture colours appear to have the combined attributes of their components, the magenta appearing a reddish blue and the cyan a bluish green. However, where the red and green overlap we get a surprising result as a mixture—namely a yellow—which appears neither to have any attributes of red nor of green. Furthermore where all three colours overlap we see a white, which is biased towards no particular hue. In other words if we add three brilliant colours together they disappear giving white. If we now vary the intensities of the three projectors we find we can produce a match to most colours where all the three colour patches overlap. We can even match a monochromatic spectral yellow with a suitable mixture of red and green lights. This means that the same visual sensation can be produced by quite different physical stimuli. Although Newton quoted an analogy with sound (section 4.1, p. 63), there is little parallel in this case except that we are dealing with vibrations and perception. If we sound two musical notes simultaneously we hear a chord, but we can discern that there are two notes, and a trained musician could name them. We do not hear a single note of some intermediate frequency, which we would do if the ear corresponded in action to the eye.

Considering Young's experiment further, we call the three basic components of any colour sensation, that is the red, green and blue lights, additive primary colours. There is in fact an infinite choice of primaries, but to obtain the maximum range of mixture colours one must choose a red, a green and a blue. The only restriction is that it must not be possible to match one by a mixture of the other two. In fact, a selection of three pure monochromatic spectrum colours makes equally good primaries as impure filter colours, although in practice they are rather dim.

The other relevant facts of colour vision are that a normal observer can detect over 150 different hues in the spectrum using a 2° diameter viewing field, and considerably more with a larger field. Furthermore, the whole gamut of colour perception is incredibly large, and we can in fact detect several million distinct colours.

4.5 YOUNG'S THEORY OF COLOUR VISION

It was obvious that there could not be a great multiplicity of different nerve pathways from the eye to the brain which could signal separately each observed colour. It was clear to Thomas Young that since the triplicity of colour had no foundation in the theory of light it must be a property of the eye. He postulated in 1802 that the eye analyzed each colour and signalled it to the brain along three different types of nerve, one type signalling the red content, another the green and a third, the violet. A magenta perception was assumed to result from a simultaneous stimulation of the 'red' and 'blue' nerves, a yellow from a simultaneous stimulation of the 'red' and 'green' nerves, and so on. When all three types were stimulated equally, a white was perceived.

4.6 HELMHOLTZ'S OBJECTION

Young's theory was first rejected, and then ignored for about 50 years until it was taken up almost simultaneously in about 1852 by the German physicist and physiologist Hermann von Helmholtz (1821–1894) and by the Scottish physicist James Clerk Maxwell (1831–1879). Helmholtz carried out some colour mixing experiments and came to the conclusion that an explanation of colour vision needed the postulation of more than three primary processes. The difficulty was that although most colours could be matched by an additive mixture of the three primary colours in suitable proportions, there were some colours which could not be matched in this way. For instance, if an attempt is made to match a spectral blue-green of wavelength 500 nm with a mixture of primary blue and green, the mixture will always appear too white (less saturated) than the spectral colour. Saturation is used in the sense that spectral colours are the most saturated colours (physically these are of the highest purity), whereas the addition of white to them produces pastel colours which are less saturated (or of low purity). (A neutral grey or a white are of zero saturation.) Helmholtz knew from his experiments that four or more primaries were needed to match some colours, and he naturally thought that Young's three-nerve theory was untenable.

4.7 HELMHOLTZ'S SUBSEQUENT ACCEPTANCE OF YOUNG'S THEORY

Helmholtz held these views for about ten years until he realized that the experiments could be explained on the basis of three fundamental mechanisms by assuming that they had broad and overlapping spectral sensitivities, in other words, although the red process was maximally responsive to red, it nevertheless was also sensitive to the green and blue-green parts of the spectrum. We know now that the response curves are of the type shown in Fig. 4·10. Thus a monochromatic yellow light of wavelength 585 nm will not only stimulate the red receptor, but will also cause considerable stimulation of the green receptor. Thus, although the stimuli may be pure, the response of the eye is not.

4.8 WRIGHT'S ANALOGY

Wright, of Imperial College, London, has given a metallurgical analogy which is helpful in explaining this. If we have a number of alloys each containing zinc, tin and copper in varying proportions, we can match a given sample by compounding new alloys by mixing pure zinc, tin and copper until their compositions exactly match. If however we wanted to match a given sample of brass (Cu,Zn) which contained no tin, and our highest purity copper and zinc available each contained a small proportion of tin, the brass could never be matched in composition by a mixture of them, because any such alloy would contain some tin. What can be done in this case, however, is to add a small measured amount of tin to the brass sample. It could then be matched by adding together the impure copper and zinc. Spectral colours are not impure in the same sense, but in general owing to the overlapping receptor spectral response curves, a monochromatic spectral colour will stimulate at least two types of colour receptor, thus giving an impure response.

4.9 THE YOUNG–HELMHOLTZ THEORY

When Helmholtz realized the important point that the retinal colour receptors could have overlapping spectral response curves, he unreservedly accepted Young's hypothesis. Since then the so-called Young–Helmholtz theory has been very widely accepted, has been placed on a sound mathematical basis, and has been the guiding principle for much visual research until very recently. On the other hand it has also been the object of some very savage criticism.

There were two main objections. Until recently no difference between the retinal cone receptors had been found, and there was in consequence no proven anatomical basis for the theory. Secondly, the theory was difficult to reconcile with the colour sensations actually experienced. One can distinguish at least four qualitatively different colour sensations, namely red, yellow, green and blue, and to these can be added white, making five in all. None of these taken singly can be seen to have any of the attributes of the others. So how can five distinct psychological primaries be reconciled with three physiological processes?

4.10 THE WORK OF MAXWELL

James Clerk Maxwell was one of the greatest theoretical physicists of the 19th century, in spite of his early death at the age of 48. He is best known for his electromagnetic theory of light, and for his prediction of radio waves 19 years before they were discovered by Hertz. His interests were, however, very diverse and today seem quite remarkable in an age of intensive specialization. He wrote remarkably original

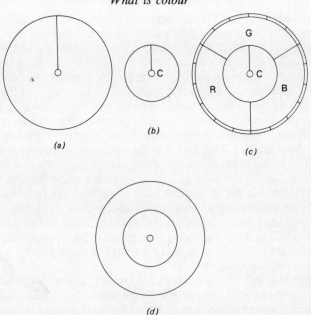

Fig. 4.7 Maxwell's coloured discs.

papers on Saturn's rings, geometrical optics, the theory of instruments and optical instruments, and the stiffness of frames, but in spite of his theoretical genius, he devoted a great deal of time to performing very elegant experiments on colour. From the first he accepted Thomas Young's theory, and although he did not develop his theoretical ideas as far as Helmholtz, he spent his time inventing accurate methods for measuring colours—methods not improved upon until comparatively recently. The elegance of some of these experiments lie in their simplicity, and these can easily be repeated at little expense. One of these was his spinning top for colour mixing, which although he used this in the simple toy form, the disc is now usually spun with a small electric motor.

Coloured papers are cut out in the form of discs which will fit on the motor spindle, and a radial slit is cut in each (Fig. 4.7(*a*)). Three large discs are cut from red, green and blue papers to form the primaries*, and a smaller disc *C* (Fig. 4.7(*b*)) is cut from a test colour which it is required to match. The larger discs are interleaved so that different angular sectors of each are exposed. The relative amount of each primary can be altered by varying the angle. This can be read off a protractor which is attached to the circumference. Maxwell found that by rapidly

* See Appendix, p. 183, for experimental details.

spinning such discs until fusion occurred, the results were identical with those found by projecting lights additively on a screen.

If, therefore, the smaller test colour disc is mounted above the primary colour discs (Fig. 4.7(c)) and the whole spun rapidly, the appearance is of a circle surrounded by an annulus (Fig. 4.7(d)). The angles of the exposed primary colour sectors are then adjusted until the circle and the annulus match. It may be necessary to add a small amount of black or a neutral grey to either to make them match in brightness as well as in colour. Maxwell used Young's colour triangle to represent his mixtures by placing the primaries R, G and B at the apices of an equilateral triangle (Fig. 4.8). The resultant colour C of any mixture of R, G, and B is found to be located at the centre of gravity of three masses at the apices, whose values are the number of degrees used of each of the primary colours. This can be verified experimentally. It also follows (and this can easily be demonstrated by the top) that the resultant of an additive mixture of two colours C_1 and C_2 is at their centre of gravity C_3, and it therefore lies on the straight line joining them. This centre of gravity law is a property of all plane colour diagrams.

In 1860 Maxwell also described his famous 'Colour Box' which was the first sophisticated instrument to be made for measuring colour. This enabled spectrum colours to be used as primaries in a most elegant way. It would not be very difficult, and would certainly be a most instructive project, to reconstruct a model of his original box. It consisted of a long wooden box (Fig. 4.9) containing two prisms P and a lens L. If

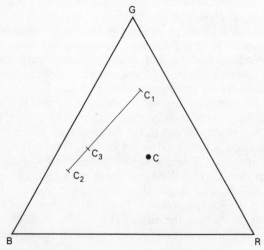

Fig. 4.8 The Young–Maxwell colour triangle.

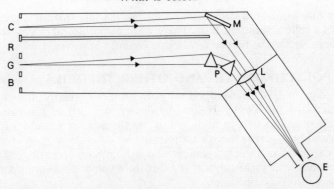

Fig. 4.9 Maxwell's colour box.

we imagine light entering at E half would pass to the mirror M and be reflected to C, and the other half would form a spectrum along the line *RGB*. If we now reverse the direction of the light and make slits at R, G and B in the red, green and blue parts of the spectrum and place our eye, at E, the light entering the slits will be recombined and we will see its additive mixture in one half of the lens. The other half will be illuminated by the light which enters at C. Thus, any colour allowed to enter at C can be matched by a suitable adjustment of the width of the slits at R, G and B, and these widths can be used as a measure of the amounts of the colours. In fact Maxwell used a white surface illuminated by sunlight, which also illuminated C. He found the proportions of the primaries needed to match all the colours of the spectrum in turn, by the application of normal algebra to the amounts of his three primaries. He used the notation x, y and z for the primaries and this is still used today in colour equations.

4.11 HERING'S OPPONENT COLOUR THEORY

In 1870 the German physiologist, Ewald Hering (1834–1918), formulated the so-called opponent theory of colour vision. He was not only impressed by the existence of the five psychological sensations described above, but also by the fact that they seemed to operate in opposing pairs, to both complement and oppose one another. Red and green seem to oppose, and do not blend. There is no reddish green colour in the same way that there is bluish green or reddish yellow. Similarly, yellow and blue seem antagonistic, and yellowish blue is never seen. He also assumed there must be a third black-white mechanism. This theory certainly explained the existence of the five psychological primaries and the complementarity of negative after-images. For example, the after-image of a bright red stimulus seen against a white surface is green, and that of a yellow stimulus, blue. There was no anatomical or physiological justification for the theory until quite recently however.

4.12 POLYCHROMATIC AND OTHER THEORIES

Hartridge in 1947 developed a polychromatic theory in which he assumed that the Young–Helmholtz theory was only an approximation to the truth, and that in addition to the three main receptors (orange, green and blue-green) there were four or five others operating in a secondary capacity, including a yellow and a blue pair operating as a unit. There was never great support for this theory and it was considered too *ad hoc* and too complicated.

There were a number of other theories proposed including Granit's dominator and modulator theory in which one type of receptor was assumed to signal only luminance whilst others supplied the colour information. There were, however, difficulties with all these theories.

4.13 MODERN RESEARCH AND MODERN THEORIES

It is interesting how our understanding progresses by fits and starts, and after Hering there was a gap of nearly a hundred years until a better understanding of colour vision was possible.

From experiments on colour mixing with both normal and colour defective observers, we have for some time had psychophysical evidence that there are three types of colour receptors in the human retina. Thomson and Wright in 1953 published curves representing the fundamental spectral responses of the three-colour mechanisms (Fig. 4.10). (Note that the 'red' curve in fact peaks in the yellow, but nevertheless has considerable red sensitivity.) Other evidence came from reflection studies of the retinal pigments *in situ* by Rushton. It was not until 1964 that direct confirmation was produced, however. This was done by two groups in America, by Marks, Dobelle and MacNichol who worked on goldfish, monkey and human retinae, and by Brown and Wald on human retinae. One difficulty is that the retinal material had to be prepared within a few hours of death. These groups nevertheless carried out the most delicate microspectrophotometry of single cone receptors, and found three types of cone which absorb light in different regions of the spectrum in a very similar way to the curves shown in Fig. 4.10. This suggests on the face of it a confirmation of the Young–Helmholtz theory. However, before this, Motokawa and Svaetichin in 1953 working independently on fish retinae, found that there were certain cells which respond by slow potential changes dependent on the strength of the stimulus, and that there were other cells in which the potential became more negative as the light was made blue, but that it changed sign and became positive for yellow light

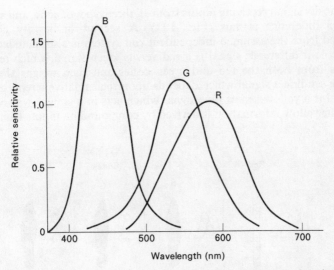

Fig. 4.10 The fundamental spectral responses of the three colour mechanisms in the eye
(after Thomson and Wright).

of longer wavelength. Other cells were also found which would reverse in potential for a red to green change in wavelength. This was followed in 1958 by De Valois and his collaborators who worked with the macaque monkey which has a colour vision mechanism almost identical with that of man. Furthermore, it is a very intelligent animal and quite amenable to training. In some of these experiments the electrical activity was recorded in the lateral geniculate body in the base of the brain. This may reflect simply the responses from the retinal ganglion cells, or may be the result of these modified by further activity in the lateral geniculate.

These lateral geniculate cells fire spike potentials spontaneously, and some either increase or decrease their rate of firing depending upon whether red or green light falls on them. Others likewise respond to yellow versus blue light. These are called opponent cells since they assess the relative strengths of opposing pairs of colours. Another type called non-opponent cells seem to signal just luminosity information.

This recent exciting research has at last given fairly firm clues to the probable action of the human colour mechanism. At the receptors the light appears to be received by three different types of cones as postulated by the Young–Helmholtz theory, and these are sensitive to the red, green and blue regions of the spectrum respectively. However, the outputs from them appear to be changed into spike discharges and coded before they are transmitted to the brain. This coded information is sent as a luminosity signal receiving inputs from all three types of cone, and as two-colour difference signals (Fig. 4.11). A second luminosity channel derived from the assumed independent rod system is also included. The first colour difference signal is a red versus green signal, which receives inputs from both the red and green cones and then weighs them up before sending a signal which depends upon their relative strengths. The second is a yellow versus blue signal which acts in a similar way, except that the yellow information is derived by compounding inputs from both

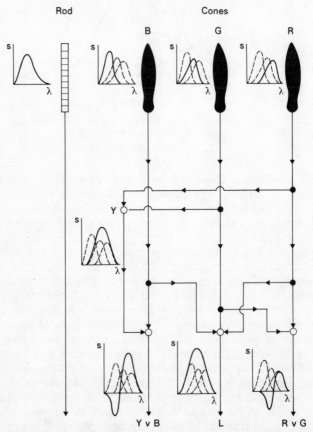

Fig. 4.11 The coding of information in the retina into luminance and colour difference signals. The relative sensitivities of each mechanism [rod, blue (B), green (G) and red (R) cones, the luminosity (L) and colour difference (YvB) and (RvG) mechanisms] are shown at each stage. The broken lines refer to the single cone responses, and these are similar to those of Fig. 4.10. (see also section 10.2)

the red and the green cones. Thus it appears that both the Young–Helmholtz and the Hering theories are valid, the first at the receptor level and the second at a later stage in the retina after the signals from the receptors have been coded. It is interesting that after all the heated controversy modern research should have declared both contestants as joint winners!

It is now possible to see with hindsight how the muddle occurred. Psychophysical colour mixing experiments naturally gave results directly related to the photochemistry of the cone receptor cells. On the other hand, the brain perceives only by receiving and analyzing the coded luminance and colour difference signals. Thus from a psychological perception stand-point colour vision appears to be an opponent process.

It is of course natural to ask at this stage why this very complicated system has evolved, whereas the pure Young hypothesis if true could be mediated by a much simpler anatomy and physiology. It is most interesting, however, that television engineers when designing a system to transmit and receive colour pictures arrived at a very similar method. They in fact analyze the picture into three basic colour components but then transmit the information as two colour difference signals and a luminance signal. There are several technical and scientific reasons why this method is preferable, and this subject will be discussed further when dealing with colour television in section 5.10, p. 90.

5
Producing colour

5.1 COLOURED LIGHTS

There are only three different ways of producing colour. The first is by choosing a light source which emits only light of a single or a few wavelengths. Such sources are usually electric discharge lamps. Examples are sodium vapour lamps used in streetlighting, which emit an almost monochromatic and brilliant yellow light, and also the mercury vapour lamps which emit a few radiations in the yellow, green, blue and violet regions of the spectrum. They are also used for streetlighting; the light appearing of a greenish-blue colour.

The modern laser is another example of a light source emitting light of a single wavelength.

The other two methods of producing colour are first the subtractive method which is most common, and secondly the additive method. These two methods will be dealt with in turn.

5.2 THE SUBTRACTIVE METHOD OF PRODUCING COLOUR

In this method we begin with a source of white light, such as a heated tungsten filament, which contains all the coloured radiation throughout the spectrum. By some means we take away (or subtract) those colours which we do not want. One common method of subtracting the colours is by passing the white light through a coloured glass or coloured gelatin film containing suitable dyes. These have the property of absorbing some colours strongly, whilst transmitting others with little absorption. This process which is the result of the molecular structure is called selective absorption. For instance a piece of red glass (Fig. 5.1) absorbs all the colours except red from white light. (For simplicity in these diagrams, white light is shown as consisting of just red green and blue light, but of course, all the colours of the spectrum are present.) Similarly a piece of purple glass (Fig. 5.2) absorbs all colours except red and blue.

A more common method of subtracting the unwanted colours from white light is by the use of pigments, and coloured materials such as plastics, which reflect light selectively. As shown in Fig. 5.3, a saturated green paper absorbs all the colours from white light except green which it

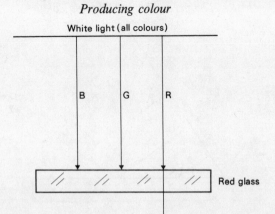

Fig. 5.1 Colour produced by red glass.
For simplicity in Figures 5.1 to 5.5 white light is shown as red, green and blue, although all colours are present.

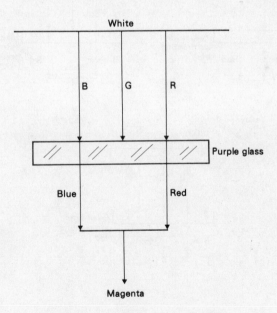

Fig. 5.2 Colour produced by purple glass.

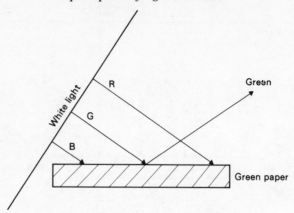

Fig. 5.3 Colour produced by green paper.

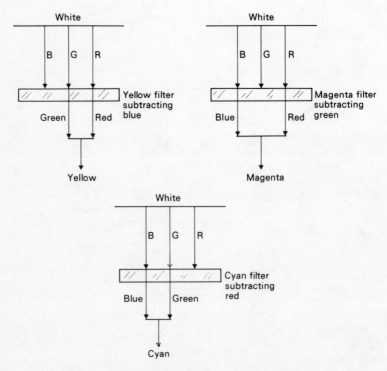

Fig. 5.4 Colours produced by yellow, magenta and cyan filters.

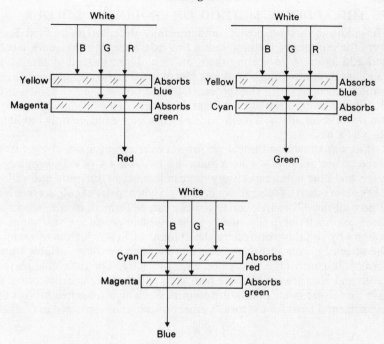

Fig. 5.5 The combination in pairs of yellow, magenta and cyan filters.

reflects. It usually happens, however, that such a surface partially reflects most colours, but more strongly reflects one or two colours. Thus, if a surface reflects all colours, but reflects more strongly say in the green than elsewhere, this is equivalent to reflecting white light in addition to green light. Such a surface gives the appearance of a desaturated (or pastel) green.

Now if we choose three colour filters* which subtract respectively mainly blue, green and red from white light, they will appear yellow, magenta and cyan (blue-green) (Fig. 5.4). If these are used superimposed in pairs, they will produce three further colours. Thus, if the yellow and magenta filters are placed together they will absorb the blue and green light respectively leaving only the red. (Fig. 5.5). Similarly the yellow and cyan together appear green, and the cyan and magenta appear blue. All three filters together absorb all the colours, and thus appear black. These are the basic principles involved when we mix water-colour paints, assuming that there are no chemical changes.

* See Appendix, p. 183.

5.3 THE ADDITIVE METHOD OF PRODUCING COLOUR

The additive method is not fundamentally different in its final form from the subtractive method, but in this case we start from a dark screen and add lights by projecting them on to it. Thus, instead of producing red light by subtracting all colours except red from white light, we simply project red light on to the screen. (Often of course the red light is produced by placing a red filter in front of the lens of a white light projector, but it could equally well be provided by a source emitting only red light, such as a neon lamp.)

If as was shown on Plate 4 we project overlapping spots of light from three projectors as Thomas Young did (section 4.4, p. 67), giving red, green and blue light respectively, we can form cyan, magenta and yellow (the colours used as subtractive primaries) where pairs of colours overlap. Where all three colours overlap a white can be formed. In fact, where all three colours overlap (i.e. add together) we can produce a vast range of colours by suitably controlling the amount of light which is thrown on the screen by each projector. This can be done by using variable transformers to control the voltage on each projector. The three colours red, green and blue are the additive primaries and they are usually chosen to give the largest range of mixture colours. Such an arrangement forms the experimental basis for the measurement of colour as explained in Chapter 7.

5.4 THE EFFECT OF THE COLOUR OF THE ILLUMINATING LIGHT ON THE PERCEIVED COLOUR OF A SURFACE

The perceived colour of a surface depends not only upon the spectral reflecting properties of its pigments but also on the colour of the light which illuminates it. For instance, a piece of paper which absorbs green light appears magenta when illuminated with white light, since it reflects strongly in the red and blue (Fig. 5.6). If, however, the incident light is predominantly blue, with very little red, the paper will appear blue. Again if the incident light is predominantly red it will appear to be red.

Such colour changes are often seen when objects are observed first in daylight and then in the light from ordinary filament lamps. There is much less blue and more red in the light from these lamps compared with daylight. Extreme colour distortion, of course, occurs when, for example, a red bus is seen under sodium lighting, which is mostly monochro-

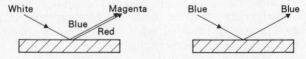

Fig. 5.6 Magenta paper seen in white and blue light.

matic yellow light. The bus looks dark brown since there is no red light to reflect, and it reflects only a little yellow light.

5.5 METAMERISM

The colour changes discussed above are gross, but smaller and more subtle changes often occur. It is possible, for instance, to have two pieces of cloth which look the same colour in daylight, but which look quite different under tungsten filament lamp lighting, even though both lights are nominally white.

The spectral reflectance curves (i.e. the percentage of light reflected in various wavelengths) of two samples of dyed cloth *A* and *B* are shown in Fig. 5.7. These both look green in daylight and exactly match in colour. This is because the reflected light, although it is of different spectral constitution, is added by the eye to give the same sensation of colour. However, when the two samples are placed in tungsten lighting, the greater red reflectance of sample *A* causes it to look brown, and the other sample to still appear green. This is called metamerism, and it can cause some embarrassment. For instance, if a patch is sewn on to a tear in a piece of clothing, it could match exactly in colour at night-time under ordinary filament lamp lighting. However, if it contains a slightly different dye from the main cloth, the patch might be very conspicuous in daylight. The only safe way of guarding against this is to use dyes which have exactly the same spectral reflection curves. Only these will exactly match in colour under all different illuminants.

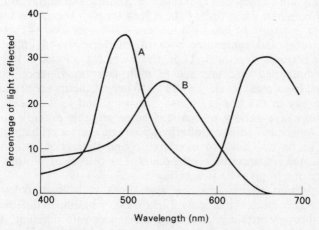

Fig. 5.7 Spectral reflection curves of dyed cloths which match in colour in daylight but not in tungsten light (after Wright).

5.6 COLOURS PRODUCED BY INTERFERENCE, DIFFRACTION, POLARIZATION AND SCATTERING

Although it was stated above that there were only three basic ways of producing colour, nevertheless a number of interesting different physical phenomena can result in the formation of colours.

Interference colours seen in thin films with white light are well known, and in these cases the colours are subtracted from white light by destructive interference. Examples of this are Newton's rings, colours of soap bubbles, and the colours seen in oily puddles.

Interference filters are now widely used in place of glass or gelatin filters. They have the advantage that they pass a very narrow band of wavelengths and thus give very pure colours, and they also have a high light transmission at these wavelengths. These colour filters are in effect Fabry–Pérot interferometers with a fixed spacing. They are made by evaporating a thin film of silver on to glass, followed by a thin layer of transparent cryolite and then finally another silver film. The cryolite acts as a spacer to keep the silver films a very small fixed distance apart. If the cryolite layer is 370 nm thick there will be maximum transmission in the green region of the spectrum at 500 nm. Radiation will also be transmitted at 1000 nm and at 333 nm, but the first is in the infra-red and the second in the ultra-violet regions of the spectrum. It is interesting that as Newton said 'vulgar People' would say that these filters are coloured, but they are in fact constructed from colourless materials, that is from materials containing no pigments. Brilliant oxide films, which appear very highly coloured by interference, can also be produced on the surfaces of metals such as tantalum by electrolysis.

Diffraction of light can also result in colours, and diffraction gratings appear coloured when white light is seen through them. It is interesting too that in nature brilliant colours are sometimes seen as the result of interference, and sometimes both of interference and diffraction, in regular periodic structures. Thus, the iridescent colours of the feathers of the humming bird are due to multi-layer interference from light reflected from platelets stacked up to 15 deep. Colours are produced in a similar way in the 'eye' of a peacock wing, and in the metallic green reflections of a variety of beatle. One remarkable example too is the South American Morpho Butterfly whose wings are a brilliant blue. This seems to be due both to interference and diffraction caused by the complicated rib structure on the scales. The colours of Mother of Pearl also are due to periodic structuring.

Polarization phenomena can also result in brilliant colours from apparently colourless materials. Light waves vibrating in different planes travel through certain crystals and substances with different velocities. This is called birefringence. It results in a delay in transmission of certain frequencies, which leads to cancellation of certain colours by interference. Thus, discrete light frequencies can be removed from white light giving rise to brilliant colours. This can easily be demonstrated using ordinary transparent self-adhesive acetate tape ('*Sellotape*'), which is birefringent by virtue of having been stretched longitudinally in the manufacturing process. If strips of tape are stuck on to a glass slide so that different parts have a different number of layers, all aligned horizontally, and this is placed between 'crossed' polaroid sheets (i.e. in a simple 'crossed' polariscope) and viewed with a white light source, brilliant colours will be seen. Table 5.1 gives the colours seen in various thicknesses of tape. It is interesting then to view these in a 'parallel' polariscope, that is, with the polaroid sheets parallel. The colours are the complementary colours of those originally seen (i.e. the colours needed to be added to the original ones to produce white light).

Fascinating colour patterns can be produced by cutting out different shapes and overlapping them, and it seems even more remarkable that these patterns are quite colourless when removed from the polariscope and viewed in ordinary light.

Natural crystals and mica produce some beautiful patterns when viewed in a polariscope, as does glass and plastic to which mechanical stress has been applied.

The fact that small particles can scatter light can also result in some beautiful colours. The light-scattering power of myriads of particles which are large compared with the wavelength of light, such as exist in clouds and in fog, is independent of wavelength. However, when the particles are small compared with the wavelength (λ) of the light, the scattering power according to Rayleigh's law is proportional to $1/\lambda^4$. This means that the amount of light scattered at the blue end of the spectrum (short λ) is about ten times that at the red end (long λ). This explains the blue appearance of tobacco smoke. It also explains the blue colour of the clear sky, and in this case the scattering is almost entirely

TABLE 5.1

Number of thicknesses	Colours seen in 'crossed' polariscope	Colours seen in 'parallel' polariscope
1	Light yellow	Blue
2	Cyan	Orange
3	Magenta	Green
4	Light green	Purple
5	Dark green	Magenta
6	Rose	Green
7	Blue-green	Rose

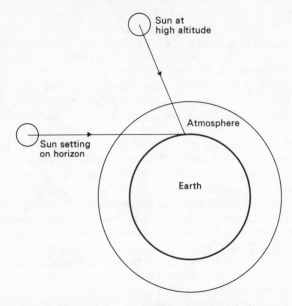

Fig. 5.8 Why the setting sun appears red.

by the actual molecules of air. If there were no scattering the sky would appear black, except for the sun, stars, moon and planets. The increasing redness of the setting sun is due to the fact that the light from the sun has to pass through a greater thickness of atmosphere, and thus more air molecules are encountered and more of the blue light is scattered away from the line of sight (Fig. 5.8).

5.7 COLOUR REPRODUCTION
Our lives are lived in a coloured environment and much thought and planning is given to making this appropriate and pleasant. Results are achieved by using either self-coloured materials such as glass, plastic, linoleum, etc. or by applying pigments in the form of paints, emulsions, varnishes, etc.

However, if we require to communicate in colour we use the more sophisticated methods of colour photography, colour printing, and colour television. These are methods of reproducing colour. They do not aim at replicating an exact physical equivalent to the object, that is an exactly similar spectral energy distribution, but in achieving an equivalent perception. Thus, a saturated near-monochromatic yellow may be reproduced by an additive mixture of red and green lights which appear to be the same yellow colour. The reproductions are therefore often metameric matches to the real objects. The basic fact of visual trichromacy is almost always used when reproducing colours.

5.8 COLOUR PHOTOGRAPHY

The first colour photograph was made by Maxwell in 1861. Curiously enough he did not set out to produce a coloured picture, but he wanted to illustrate the three-colour basis of human vision. He took three separate black and white photographs of a scene, one through a red filter, one through a green filter, and the third through a blue filter. He then made three black and white positive lantern slides. These were simultaneously projected on to a screem from three projectors in front of each of which were placed the same three filters in their correct respective positions. Physically there were only three colours on the screen, but subjectively the eye sees a whole range of colours, and the results are most realistic. In modern colour reproduction whether by colour photography, colour printing or colour television, the principle of Maxwell's method is still used almost universally, although the actual details sometimes may not immediately appear to be directly related.

One of the disadvantages of Maxwell's method lies in its complexity, and also the fact that since the three-colour photographs have to be taken *successively*, only stationary objects can be photographed. This could be overcome by using three separate cameras with matched lenses, but this complicates the apparatus and increases the expense. Another disadvantage of this additive method is that the three colour filters must be saturated, and this inevitably means that they will have a low light transmission—the result is a dim final reconstructed image. Subjectively the eye perceives a much greater range of colours when looking at bright images, so a dim image leads to a smaller range of reproduced colours.

Although colour photography is now almost entirely based on the subtractive principle, systems employing the additive method were formerly very widely used. They depended on the so-called mosaic method. If a very fine mesh of red, green, and blue squares is viewed from a sufficient distance, the individual coloured patches are not seen separately by the eye; what is seen is a uniform colour. This is the result of the addition of the varying proportions of light from all the three types of coloured squares. If more light passes through the red and green than the blue squares, a yellow colour will be seen, similar in colour to that which would appear if red and green lights were projected simultaneously onto a screen.

In this type of colour photography a coloured mesh as described above is placed in contact with a photographic plate in a camera. This is then exposed. The image is developed and reversed so that the dark parts become light and the light parts dark, and the reversed image is then viewed with the coloured mesh still in contact. Colours will be seen corresponding to those in the original scene. This method has the great advantage that only one photograph is taken, and this can either be viewed without a projector, or alternatively projected by a single projector. The Autochrome plate was based on this principle, but instead of a coloured mesh, a random mixture of very small red, green, and blue dyed starch grains was used. This process was used commercially from 1907 to 1930. The Agfacolor process was similar and used a random mosaic of stained resin grains. Several successful processes were evolved using ruled screens, one being Dufaycolor in which the mesh had a million squares to the square inch. The main disadvantage of these methods was their comparatively low light transmission and consequent rather dim pictures, and also the lack of fine definition caused by using a mosaic.

The various subtractive processes have now superseded these, and give much brighter photographs. They depend on the production of three superimposed dye images, using the three subtractive primaries as dyes. The depth of the cyan dye controls the amount of red left in the image, the yellow dye controls the blue content, and the magenta dye the green content. All that is necessary is to control the concentrations of these three dyes independently at each point of the picture. One obvious way of doing this, perhaps, would be to take three separate photographs through red, green, and blue filters as in Maxwell's method, and then to make dye-image positive transparencies from the negatives. The red negative is used to produce the cyan image, the green to produce the magenta one, and the blue to produce the yellow. The three transparencies are then exactly superimposed and viewed with white light. Plate 5 shows the appearance of the three separation positives and a black impression, and all of these superimposed to give the final colour picture.

However, it is normally quite impractical to take three photographs, or alternatively to have three similar cameras, so modern methods use three emulsions on top of one another to record the three images. This subtractive colour process is known as the *integral tripack*. Photographic emulsions are mainly sensitive only to blue light, but their sensitivity can be extended to include green and red light by the addition of sensitizing dyes. The top layer (Fig. 5.9) is an ordinary blue-sensitive emulsion, and it records the blue image. This is followed by a yellow (minus blue) filter which renders ineffective the blue sensitivity of the two emulsions underneath. The second emulsion is sensitive only to blue and green, and thus records only the green image. The bottom emulsion is sensitive only to blue and red light, and thus records only the red image. Thus, at a single exposure the blue, green, and red images are recorded in depth in

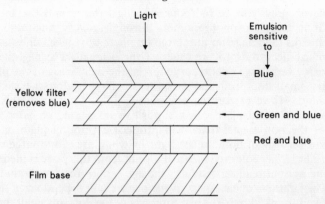

Fig. 5.9 An integral tripack colour film.

the same film. These three layers are then changed into yellow, magenta, and cyan images by a complicated process of colour development which will not be described here. The result is a single colour photograph with the separate layers completely in register. Brilliant colour transparencies and prints can be produced in this way since the subtractive dyes have fairly broad spectral absorption curves instead of the narrow transmission curves of the filters in the additive processes.

5.9 COLOUR PRINTING

Full colour printing is done by printing three superimposed subtractive images in cyan, magenta, and yellow inks in a similar way to the method of subtractive colour photography (already described in section 5.8 above. Colour printing is done mainly by three processes: namely letterpress, lithography, and gravure. Letterpress is very widely used for newspapers, books, and magazines, and is the method which will be described. As its name implies it was originally used for the printing of letters. The parts required to be printed are raised in relief above the surrounding areas on a metal plate. The non-printing areas are removed by manual, chemical, mechanical, or electrical means. The parts left in relief, the printing areas, are coated with printing ink by passing an inking roller over the plates. When the plate is then pressed firmly into contact with a piece of paper, the ink is transferred, giving the required pattern on the paper. With simple letterpress, therefore, it is possible to print in one density only. We thus have either black or white, or a colour or white, but we cannot, for example, print a shade of grey between black and white. It is therefore useless for ordinary photographs where many intermediate shades between black and white are required.

However, greys can be reproduced using letterpress by an ingenious trick. It is done by photographing through a very fine-mesh screen which breaks the image up into large numbers of very small dots, which vary in size according to the tones of the original. The negatives thus made carry very large dots recorded from the light areas of the original, and very small dots from the dark areas, which is of course the wrong way round. However, subsequent photographic transfer of this image onto another metal plate yields a positive image, which after etching away of the non-image areas, will produce a printing plate on which there are very small dots on the light or white areas and large dots on the dark or highly coloured areas. If a print from this plate is then viewed from the normal reading distance, the eye cannot resolve the individual dots. The lightness of each part of the picture will depend upon the ratio of uninked to inked paper. Therefore, where the dots are small the paper will be light, and where they are large it will be dark or strongly coloured, with of course all shades and saturations of colour in between. This is called half-tone printing, because it is possible to produce these intermediate shades. The dots can be seen using a hand magnifier, and they are especially easy to see on a newspaper photograph. This is because the screen used for newspapers is fairly coarse with 22, 26 or 33 dots and spaces per cm, whereas in illustrations in books and reference works the screens used are much finer with 47, 52, 59 or 79 dots and spaces per cm. The finer the screen the greater is the amount of detail which can be printed, and since the eye is less able to see the dot structure, the result is smoother and more pleasing. Plate 6 (*top*) shows part of a positive half-tone photograph highly magnified.

It is a short step to produce three dot images by using suitable screens and to print them on top of one another in cyan, magenta, and yellow inks. Often a fourth image is superimposed using black ink. Plate 5 has been printed in this way. The four separate component images are shown. The black is used mainly because of certain deficiencies in the printing inks, since a good black is not always produced when just three coloured images are superimposed. Plate 6 (*bottom*) shows a small area of a letterpress colour photograph greatly enlarged. It will be seen that such an image is not formed by a purely subtractive process, but that it has some additive features as well. This is because in certain areas of the image the random cyan, magenta, and yellow dots are not properly superimposed and are printed separately.

5.10 COLOUR TELEVISION

5.10.1 The field sequential system

The simplest method of producing television in colour is to use the field sequential system. Red, green, and blue filters are rotated rapidly in turn in front of a single black and white television camera, and similar filters are rotated in synchronism with these over the screen of a black and white television receiver connected to the camera. This method has been used for closed-circuit television, but it is obviously inconvenient to have rotating filters in front of a receiver screen, and the frequency of alternation has to be quite high, at something like 150 red, green, and blue fields per second to eliminate flicker. The eye of course operates additively in this case when the fields are changed sequentially at a speed too high for detection. Another disadvantage is that the system is not compatible by which is meant that pictures transmitted by such a system could not be received as black and white pictures on existing sets.

When we now come on to discuss modern systems it is interesting to point out that although the mosaic system of colour reproduction is now obsolete for colour photography, it has come into its own in colour television, where most receivers now contain mosaic dot cathode-ray tubes. In these systems the camera (Fig. 5.10) consists of a single lens and beam-splitting mirrors. Three camera tubes are used with red, green, and blue interference filters, to produce the red, green, and blue signals as the image is scanned line by line. However, separate red, green, and blue signals (ER, EG and EB) are not in fact transmitted by the television stations, since if they were the broadcasting of colour pictures would require three times as much frequency space (called *band-width*) as normal black and white pictures. What is done is to transmit a luminance signal (EL) which indicates the brightness of each part of the picture, together with two colour difference (or chrominance) signals such as (ER—EL) and (EB—EL). It is found that if the luminance signal carries

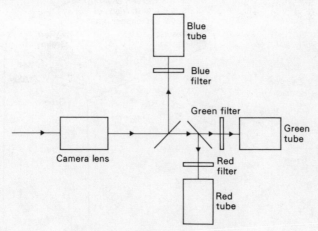

Fig. 5.10 A colour television camera.

an information content of 100, the two chrominance signals need only carry an information content of 25 each for a good colour picture to be produced. This means the total information needed to be broadcast is only 150, compared with 300 if the red, green, and blue signals were sent separately. This results in a considerable saving in band-width. It also has the added advantage of being compatible, that is a black and white receiver can operate on the luminance signals (the chrominance signals being ignored) and thus reproduce good black and white pictures.

A colour television receiver operates on a red, green, and blue additive principle, so the red, green, and blue signals given by the three camera tubes have to be recovered from the luminance and chrominance signals by electrical circuits in the receiver.

These signals are then fed into three electron guns in the receiver tube, which fire three electron beams whose strengths vary with the three signals (Fig. 5.11). The screen of the tube is covered with triads of minute red, green, and blue phosphor dots. Before the electron beams reach the screen they pass through a shadow mask, which is a metal plate perforated with a large number of small holes. This enables the electrons fired from the gun activated by the red signals to fall only on the red-glowing phosphor dots, those from the 'green' gun to fall on the green dots, and those from the 'blue' gun to fall on the blue dots. There are about a million separate dots on a normal screen, but they cannot be seen

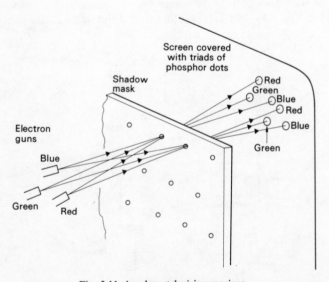

Fig. 5.11 A colour television receiver.

at the normal viewing distance. (A close inspection of a working screen with a hand magnifier will easily reveal the separate dots.) The eye therefore sums up the simultaneous effect of the three types of dots in the same way as in the old mosaic method of colour photography.

It is of interest to note that as we saw in Chapter 4 the eye itself processes the visual information falling on the retina in a similar way to this method of colour television, and produces red, green, and blue signals from the retinal receptors, but then transmits luminance and two colour difference signals to the brain.

When we are watching a colour television screen the colour in the original scene is first being analyzed into its red, green, and blue components by the camera, then transformed into luminance and chrominance signals before transmission. On reception at a distant location, the signals are converted back into red, green, and blue signals and presented to the eye on an unresolved mosaic screen. The image of this is then formed on the retina by the eye's optical system and the red, green, and blue information is added together in the retina. The retinal neurons thus process the information back again into luminance and colour difference signals which they send to the brain. After all these complicated transformations it is perhaps not surprising that on occasions the result is a not too faithful reproduction of the original scene.

6
Colour appearance systems

6.1 SIMPLE COLOUR SPECIFICATION BY APPEARANCE

As we implied in Chapter 4 a source of light containing a band of spectral energies is not perceived as being composed of individual wavelengths but is seen as a single colour. Under low stimulus levels (i.e. for scotopic vision) all wavelengths have the same visual effect and we can only perceive degrees of light and shade. In this case a single receptor mechanism is functioning and the colour appearance of an object can be specified by a single scale of luminosity. With higher stimulus levels (i.e. photopic vision) colour perception is more complex and colour specification requires at least three perceptual dimensions. This minimum number is no doubt related to the response of the three distinct types of cone receptors present in the retina. Changes in object size, texture, background colour and in the adaptation state of the observer can all affect the colour appearance and in general we cannot expect always to be restricted to three dimensions of colour perception. However, under many common visual conditions a colour can be described adequately by just three attributes of colour. This is one example of the phenomenon of colour constancy.

6.2 COLOUR CONSTANCY

When an object is viewed under different lighting conditions, perhaps first in an artificially illuminated interior and then under daylight, its colour appearance often remains surprisingly constant. This phenomenon, called colour constancy, occurs so frequently even for relatively large changes in the level and colour of the illuminant and the background that it soon ceases to be a surprise. Most visual scenes are complex, being made up of many surfaces of varying chromaticity and reflectance and any one of these may be viewed against a different arrangement of the other surfaces. The average background colour (i.e. the composite effect of all the coloured elements in the scene) will generally approximate to an achromatic surround of low to moderate reflectance. If we restrict ourselves to the appearance of objects subtending a few degrees of angular subtense at the eye, the field size, background chromaticity and relative luminance of the object to the background remains approximately constant. Changes in the illuminant level do not influence these factors and small variations in the illuminant are largely counterbalanced by chromatic adaptation effects of the eye. This invariance in the physical conditions seems to result in constancy of colour perception. This phenomenon is certainly only approximate and many exceptions can be exhibited (see for example the contrast effects in Plate 8 (*top*) but it does allow us to describe colour appearance in a relatively simple manner. For example, if the chromaticity (see section 7.2, p. 114) and relative luminance of an object are known a useful prediction of its hue, saturation and luminosity can be made.* It is only because of this constancy that a colour appearance specification (that is a psychological rather than a psychophysical specification) has such a wide application.

A number of colour appearance systems have been developed. These can be divided into systems based on (a) an empirical and subjective arrangement of a large number of coloured surfaces; (b) the empirical study of chromaticity differences; and (c) theories of vision which attempt to relate what we see to the psychophysical stimuli.

6.3 COLOUR ATLASES AND CHARTS

The colour atlas is the simplest way of presenting an arrangement of coloured surfaces (often called samples or chips). The appearance of each sample is assumed to be adequately described by a limited number of its psychological attributes, e.g. its hue and the magnitude of its saturation and its luminosity. (The term luminosity is often replaced by lightness when surfaces rather than coloured sources are being discussed.) Each page of the atlas contains a two dimensional display of the samples arranged to show the perceptual variation of these attributes taken two at a time (Plate 7 (*top*)). One disadvantage of the colour atlas is that the samples are likely to become dirty and discoloured with use. There is also a practical limitation to the number of samples which can be included. Consequently there are gaps and missing colours; in particular some of the highly saturated colours may be impossible to produce as surfaces. On the other hand the colours can actually be seen by the user.

6.3.1 The Munsell Book of Color

In 1905 Munsell, an artist, endeavoured to produce a colour atlas which would assist in the teaching of the concepts of colour to students of painting and related fields. The resulting *Munsell Book of Color* has since been extensively revised and now contains over 1200 colour samples both with matt and glossy finishes; it is perhaps the simplest subjectively based colour system and the most widely used at the present

* See Glossary of Terms, p. 180.

time. Its psychological and psychophysical properties have been extensively documented and it can be considered as an excellent example of a means of specifying colour appearance.

When used correctly each coloured sample is viewed under standard observation conditions, namely against a uniform grey background of 20 per cent reflectance and with both chip and background illuminated by a source similar to the colour of daylight. Colour constancy can therefore be expected to occur. The colour appearance of each chip is assumed to be uniquely described by its hue, saturation (called Munsell chroma) and luminosity or lightness (called Munsell Value). The chips are arranged in two dimensional arrays of chroma and Value such that adjacent chips increase by equal perceptual intervals of saturation or luminosity (Fig. 6.2 & Plate 7 (*top*)). It is convenient to describe first the idealized Munsell system and to stats its limitations afterwards.

6.3.2 Munsell hue

Any coloured surface appears to have an attribute related to one of the spectral colours or to a highly saturated purple. This is called its Munsell hue. There are five principal hues: blue, green, yellow, red, and purple (designated by B, G, Y, R, and P) and five intermediate hues (BG, GY, YR, RP, and PB). These ten hues can be arranged in order in the form of a circle (or hue circuit) such that the apparent hue difference between neighbouring surfaces is constant. All usefully discernible hues are specified by subdividing each segment into ten further hues giving 100 in all (Fig. 6.1(*a*) & (*d*)). The magnitude of the colour difference between any two hues is indicated by their angular separation on the hue circuit or by their difference in hue number. In practice the hue numbers convey little meaning to the inexperienced subject and therefore an alternative specification is used. Each hue is specified by a number between 1 and 10 followed by the designation of the ten major hues (for example 3R, 7BG, etc.). By convention the number 5 refers to a match with the corresponding major hue, e.g. 8BG refers to a hue 3 steps from pure blue green (5BG) towards pure blue.

6.3.3 Munsell Value

Each surface has an attribute which relates to the luminosity of the achromatic series of colours from black through shades of grey to white. If black and white surfaces are given Munsell Values 1 and 10 respectively intermediate greys of values 2 to 9 can be chosen which appear to be separated by equal perceptual intervals of luminosity (Fig. 6.1(*b*)). Chromatic chips have the same Value as the achromatic chip of the same luminosity. Since the illuminance of each chip is the same the Values of different coloured surfaces will be identical for similar luminances or reflectances.

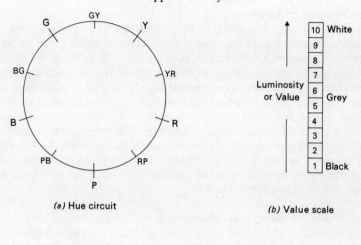

(a) Hue circuit

(b) Value scale

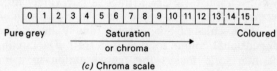

(c) Chroma scale

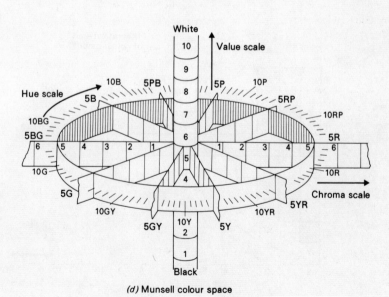

(d) Munsell colour space

Fig. 6.1 The Munsell colour system.

6.3.4 Munsell chroma

Any coloured surface appears to be a mixture of a grey and a pure hue. Colours of greater apparent hue content are said to be more saturated or possess a higher Munsell chroma. A set of surfaces of the same Value and hue can be arranged in order of chroma. In the colour atlas they are separated by equal chroma steps and are designated by chroma values increasing from 0 for grey up to perhaps 16 or 18 for the most saturated surface (Fig. 6.1(*c*)). The upper limit of the chroma number will depend on the Value and hue under consideration. There are certain theoretical limits to the chroma which can be obtained with surfaces (the MacAdam limits shown in Fig. 6.2). For example high chromas and high Values are mutually exclusive. There are also practical limits due to the chemical properties of pigment mixtures.

6.3.5 Munsell notation and colour space

Every surface is perceptually specified by the hue, Value, and chroma symbols. By convention these are written as hue/Value/chroma. Hence 5BG/6/9 specifies a Blue-Green of medium Value and high chroma.

The three colour scales of hue, chroma, and Value can be arranged as a three-dimensional colour solid (Fig. 6.1(*d*)). The circular arrangement of the hue scale makes the representation in cylindrical co-

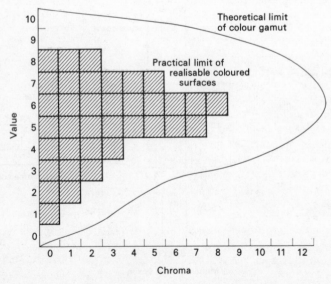

Fig. 6.2 A schematic representation of a page of constant hue from the Munsell Atlas.

ɔrdinates most suitable. An arrangement of the samples taken from a page of constant hue of the Color Atlas is shown in Plate 7 (*top*).

6.3.6 Limitations and applications of the Munsell system

Numerous experiments have attempted to test the adequacy of the spacing of the three Munsell scales. The Value and chroma scales seem to be a useful approximation to equal perceptual interval scales. However, the assumption that there are five principal hues and that these appear to be equally spaced can be questioned. Some psychologists hold the view that there are only four principal hues red, yellow, green, and blue, purple being a blue-red perception. The assumption that Value only depends on relative luminance or reflectance for all hues and chromas is only approximately correct. High chroma surfaces of certain hues are seen by most subjects to be of higher luminosity than low chroma surfaces of the same Value (the Helmholtz–Kohlrausch effect, see also section 8.4, p. 135).

The colour differences corresponding to unit increments of hue, Value and chroma are not perceptually equal. To a first approximation a unit increment of Value is easily perceived and perceptually equivalent in magnitude to a hue change of about 3 steps or a chroma change of 2 steps. This must be taken into account when considering the subjective colour difference between two surfaces of perhaps different hue, Value, and chroma especially if one attempts to relate this to their separation in three-dimensional Munsell colour space. A further difficulty arises when some such meaning is attached to this spatial separation. Several experiments have shown that a colour appearance cannot be related precisely to a three-dimensional Euclidian space. Consequently there is no simple exact relationship between the separation of two colours in Munsell space (Fig. 6.1) and the magnitude of their perceived difference. Any further detailed discussion on this topic is outside the scope of this book. However, in many applications of colour specification the requirement is to obtain an estimate of a small colour difference between two samples. For example in paint manufacture the colour difference between two batches of a similar pigment will always be small but it is important to know whether this is within some stated tolerance limits. In such cases the colour difference can usefully be specified in terms of the hue, Value and chroma differences.

6.3.7 Other colour atlases

In the DIN* colour system developed in Germany in the 1950's colour samples are assumed to be described by their hue (DIN-Farbton), saturation (DIN-Sättigung) and relative lightness (DIN-Dunkelstufe).

---

* Deutsche Industrie-Norm, which is a German industrial standard.

The hue circuit is similar to the Munsell hue circuit only it is divided into 24 equal perceptual steps of hue. Hue in this system is assumed to depend only on the dominant wavelength of the sample, an assumption which was found to be only approximately true by the developers of the Munsell system. The major difference between the DIN and Munsell systems is in the relationship of the Value and relative lightness scales. Relative lightness is defined in the DIN system as a logarithmic function of the ratio of the sample reflectance to that of the 'optimal' colour of the same chromaticity. This is intended to produce a better psychological balance to the final colour arrangement but prevents any simple comparison with the Munsell system. The DIN saturation scale is assumed to be only dependent on the chromaticity and not to vary with relative lightness. This is also a major departure from the empirical results of the Munsell system.

The Hesselgren Colour Atlas is, like the Munsell system, based on surfaces arranged in scales of hue, saturation and luminosity. This has been used recently as a basis for the Natural Colour System. The colours of this sytem are organized according to direct estimates of the relative amounts of the six natural colours blue, green, yellow, red, black and white which appear to make up the colour. On the basis of these estimates the organization of several psychological colour attributes can be assessed.

The Colour System developed in 1917 by Ostwald, a physicist, like the Munsell system, attempts to arrange the colours of surfaces in a three-dimensional space. Fig. 6.3(*a*) shows an arrangement not unlike the Munsell colour space and the two systems have similar applications. The Ostwald system however, is not based on an empirical perceptual colour organization but on the subjective opinions of its originator. The vertical axis represents achromatic or grey surfaces from black to white. A vertical section of the solid is diamond shaped (Fig. 6.3(*b*)) the two triangular areas either side of the grey axis containing complementary colours of constant dominant wavelength called 'full' colours. This is analogous to a similar section of the Munsell solid. The organization within each area is related to the relative amounts of black content, white content and full colour content. Diagonal lines either contain surfaces of constant black content and decreasing white content, or for the complementary triangle constant white and decreasing black content. The full colour content increases radially from the central axis. The colour samples were chosen according to how they match a hypothetical set of coloured surfaces which also define the black, white and colour contents. Each hypothetical colour (Fig. 6.4) has a spectral reflectance which is constant at one of two levels bounded by complementary wavelengths. In effect the organization is based on the psychophysical variables of dominant wavelength (constant in each triangle

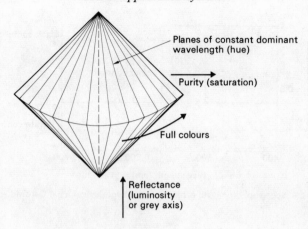

(a) Ostwald colour space

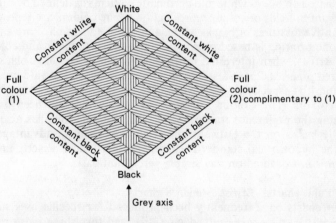

(b) Vertical section of two sets of coloured
surfaces of complimentary dominant wavelength

Fig. 6.3 The Ostwald colour system.

of Fig. 6.3(*b*)), purity (constant for a vertical series) and relative luminance or reflectance (constant for a horizontal series).* Strictly speaking
the only psychological choice is the equal spacing of the 24 full colour
surfaces of varying dominant wavelength (or effectively hue).

The Ostwald system can be used as with the other systems discussed

---

* See Glossary of Terms, p. 180.

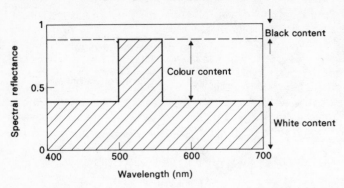

Fig. 6.4 Spectrophotometric properties of a hypothetical ideal Ostwald surface.

above in relating the colour appearance of an object to the corresponding matching Ostwald surface. It also has the advantage that its arrangement being based on black, white and chromatic contents as defined by Fig. 6.4, is relevant in fields of colour reproduction where colours are reproduced by additive mixtures of black, white, and coloured stimuli. For example, half-tone printing is based on the fusion of separate black and coloured dots. Artists, when interested in the effects of mixing white, black, and coloured pigments, have also noted advantages in the Ostwald presentation and it is especially useful when choosing harmonious (complementary) colours. The Colour Harmony Manual which has been used in assessing the preference for various colour combinations has also been usefully based on the Ostwald approach. The main disadvantages are that the hypothetical standards cannot be reproduced exactly and the system is not ordered on any simple perceptual basis.

6.3.8 Paint charts—British Standard charts
Colour charts have frequently been produced for specific uses such as in the textile, pigment, paint, dyeing, tile, and furnishing industries. In general these are limited to a range of colours suitable for a particular application, perhaps to 30 samples rather than the 1200 of the Munsell colour system. In these examples surface texture can be varied to cover a range of finishes as opposed to the two matt and glossy samples of the Munsell charts. However, to assist communication these are often also specified in the Munsell terminology. An excellent example of colour samples for specialist use is that issued by the Royal Horticultural Society. A set of 800 inexpensive coloured surfaces have been produced which are considered most suitable for the recognition and specification of the colours of flowers and plants.

National colour standards such as the British Standard (BS 266: 1955) are intended for a wider use than in one type of industry. These tend to be a compromise between the extensive coverage of the main colour systems and the charts made for specific uses. The new British Standard (BS 4800: 1972) has an organization of its 86 samples in a manner which it is thought will assist the designer in co-ordinating groups of colour combinations. A major change from the previous BS 266: 1955 is the arrangement of the coloured samples on the basis of their perceptual grey content, an attribute of colour perception which is not specifically included in most colour standards.

6.4 COLOUR APPEARANCE TERMINOLOGY

A deliberate attempt has been made in this chapter to describe colour appearance in terms of hue, saturation and luminosity at least where this is appropriate. However, in practice different terms are used in different fields of application. These vary so extensively that it is not possible to give a coherent and comprehensive list of different terms and their meanings. The term luminosity is not normally used when dealing with surfaces, lightness being the preferred term. *Brightness, brilliance* and *Value* also occur frequently. *Bright, brilliant, vivid* and *clear* are often used to describe highly saturated colours especially for objects of high reflectance. *Strength* has similar meaning in the dyeing industry both in the subjective and objective sense. *Shade* is a much overworked term taking almost any meaning although this has a serious rival in the term *tone. Tint* also takes several meanings but it most frequently refers to the results of adding different amounts of white to a chromatic pigment. Terms such as *weak, dull, deep, pastel* take fairly obvious meanings especially when applied to specific colour applications but others such as *thin, hungry* and *stingy* colours do not convey much about colour appearance. In the world of colour education there are obvious advantages in restricting the number of terms to a minimum so that hue, saturation and luminosity are the preferred, psychological terms for discussing colour appearance. They have the added advantage that they are fairly obviously related to the psychophysical correlates of dominant wavelength, purity and luminance.

6.5 COLOUR ORDER SYSTEMS BASED ON COLOUR DISCRIMINATION

A second approach to specifying colour appearance is based on the ability of a subject to discriminate a difference in colour. If a coloured field is slowly varied in hue, saturation or luminosity until a just notice-able difference (jnd) in the perception occurs the corresponding change in the psychophysical variable, i.e. the dominant wavelength, purity or luminance is known as the difference threshold. The three difference

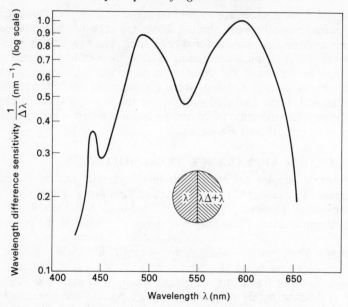

Fig. 6.5 Wavelength difference sensitivity function measured at threshold with mono-chromatic fields.

threshold functions for wavelength, purity and luminance are shown in Figs 6.5, 6.6, and 6.7. Fig. 6.5 is measured at maximum purity (i.e. for monochromatic light) and at a fixed photopic luminance, Fig. 6.6 for near white light and fixed photopic luminance, and Fig. 6.7 for white light. Other related functions could be plotted for different combinations of dominant wavelength, purity and luminance. Such a family of functions would define the psychophysical conditions for all possible colours which could just be discriminated by an observer. It has been estimated that some 7 million different colours can just be discriminated by the average observer. If such data could be obtained it would be possible theoretically to organize all these colours in a colour space where each sample was separated from its nearest neighbours by one jnd of colour difference. It is often assumed that the number of jnd steps between two colours is a measure of the perceptual colour difference. If this assumption is correct the colour solid would bear some simple relationship to the three-dimensional Munsell colour space. (See Sections 3.7 & 10.4).

Experiments have shown that a colour space based on discrimination data cannot be described precisely in three-dimensional Euclidean space but nevertheless it is thought that an approximate system would be a useful method of specifying colour and colour differences. The large number of possible colours makes it impractical to base this colour organization entirely on empirical data. Two approaches have been used to overcome this difficulty, however. The first makes use of chromaticity discrimination data together with an assumed luminosity scale and the second makes use of the theoretical concept of the line element.

6.5.1 Chromaticity differences

The CIE (*xyz*) chromaticity diagram shown in Fig. 6.8 is effectively an organization of all possible colours for a given level of luminance. There is no reason to suppose that the distance separating two points on this diagram tells us anything about the perceptual difference between the two corresponding colours. However, it is possible to transform this diagram to achieve this end. The type of transformation can be found by trial and error or other means once we know the size of one jnd of chromaticity in all regions of the diagram. It is not practical to measure all possible discriminations but a suitable chosen sample will suffice. The colour discrimination areas obtained by MacAdam are included in Fig. 6.8. These were measured by a specially designed colorimeter which

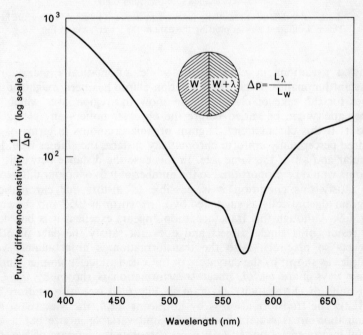

Fig. 6.6 Purity difference sensitivity function measured at threshold with white and near-white fields.

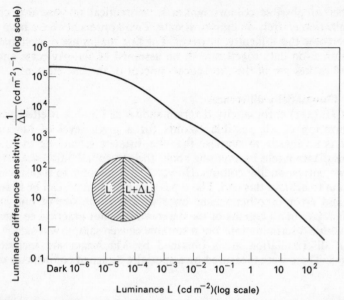

Fig. 6.7 Luminance difference sensitivity function measured at threshold with white fields. This is a different way of plotting the data of the lower curve of Fig. 3.6.

allowed variations in chromaticity while automatically maintaining constant luminance. Each discrimination ellipse has been magnified ten times for the sake of clarity. These show the region over which the chromaticity can be varied before the observer notices any change of colour. If this chromaticity diagram organized colours in terms of the desired perceptually uniform chromaticity surface, the ellipses would be circular and all of the same size. In this case the distance between two colours would be proportional to the number of jnds of colour difference. To satisfy this condition it is possible to distort our chromaticity diagram algebraically as suggested by Farnsworth in 1957 and shown in Fig. 6.9. Although this transformation appears excellent it is based on the results of a single subject and does not satisfy the data of other subjects so precisely. Also the transformation is unfortunately very complex as shown by the curvature of the x and y loci. A simpler, and in many ways more useful, linear transformation is shown by the CIE (uv) uniform chromaticity diagram in Fig. 6.10 (see also section 7.8, p. 122). The transformation is by no means ideal, the MacAdam discrimination areas remain non-circular and variable in size but it is a useful improvement on the CIE (xyz) chromaticity diagram of Fig. 6.8.

To complete our approximate uniform chromaticity space the Munsell value scale has been adopted to account for variations in discrimination as the level of the luminance (photopic) is varied. This is particularly appropriate since the Value scale based on equal intervals of the perception of luminosity has to be shown to agree with a Value scale derived from luminance discrimination data. The resulting three-dimensional colour space is known as the CIE(U^* V^* W^*) system which was recommended in 1964. Any colour difference between two colours, themselves specified by U^*, V^*, W^* and $U^* + \Delta U^*$, $V^* + \Delta V^*$ and $W^* + \Delta W^*$, is given by:

$$\Delta E = [(\Delta U^*)^2 + (\Delta V^*)^2 + (\Delta W^*)^2]^{\frac{1}{2}}.$$

6.5.2 The line element
An alternative approach is to attempt to construct a uniform colour space from our knowledge of the behaviour of the visual process. Helmholtz first derived the required information from certain assump-

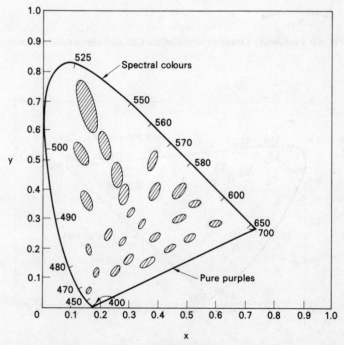

Fig. 6.8 The MacAdam discrimination ellipses plotted on the CIE (xy) chromaticity diagram. Each ellipse is a measure of the chromaticity threshold. These are shown ten times the actual size with their centres correctly positioned.

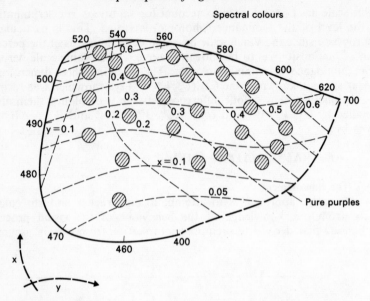

Fig. 6.9 Farnsworth's transformation of the CIE (*xy*) chromaticity diagram.

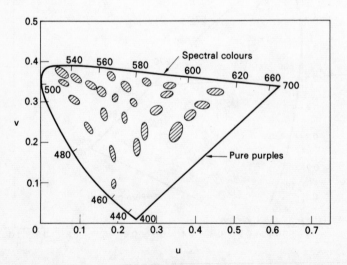

Fig. 6.10 The MacAdam discrimination ellipses plotted on the CIE (*uv*) uniform chromaticity diagram. These are shown ten times the actual size with their centres correctly positioned.

tions about the responses of the three cones and from the assumption that Weber's law of discrimination held (see section 3.8). This approach is based on the concept of the line element which is the length of a line joining two points in three-dimensional colour space. This initial study did not produce many practical results but a recent more sophisticated analysis by Stiles has renewed interest in this approach. A detailed discussion on this topic is beyond the scope of this book but an indication of its value can be seen from the fact that Stiles was able to predict many of the colour discrimination data such as the MacAdam ellipses which compare very favourably with existing empirical data. However, at the present time the theoretical approach of the line element is probably most useful in testing our understanding of the visual process. The attractive simplicity of an empirically-based uniform colour space will probably suffice for some time to come.

7
Colour measurement

7.1 THE NEED FOR COLOUR MEASUREMENT

As we shall explain below it is often necessary to make measurements of colours for industrial and other purposes. It is not easy to devise a system by means of which anything as subjective as colour can be measured. What are needed are measurements of a type which are fairly easy to make, can if possible be done objectively, which are reasonably precise, and which by their equality, will enable us to tell whether two colours will appear the same (i.e. will match one another).

It must be emphasized that colour measurement is not done primarily for perceptual reasons. The measurements of a colour only give a vague idea of its appearance, and as will be explained in Chapter 8, can be very misleading in certain situations.

One example of an industrial situation demanding colour measurement is as follows. Different parts of the body-work of a motor car are usually manufactured in different parts of a large factory, and they are brought together for the first time on the assembly line. Naturally it is most important for all these parts to appear to be of exactly the same colour so that they will match and appear to be an integral whole. Colour measurements are therefore made separately on all the component parts, and the measurements are compared. If they agree, then all the parts will match in colour. It is not necessary for these measurements to tell us precisely what colour we perceive when we look at the car. This will depend not only upon the colour measurements but also on the colour of the light, on the colour of the background, and on a number of other factors.

Further examples of industrial needs for colour measurement are found for instance in paper making. For convenience of manufacture alternate sheets in a pad of writing paper come from different rolls of paper. The eye is very sensitive to small changes of colour especially of near-white colours. Hence if accurate colour measurement and control were not done on the rolls of paper, it would be very obvious that alternate sheets were a different colour.

Often too, a pair of trousers is made by joining two pieces of cloth from different rolls. Slight differences of colour would easily be seen across the seams unless colour measurement were done and control exercised.

A great deal of colour control is necessary in the food and drink industry. We expect our butter to appear the correct shade of yellow, and our beer and soft drinks to be of the right shade from month to month. Some food such as fish can appear very unfresh if the colour is slightly greenish. This can easily be the result of using the wrong kind of lighting. Thus the measurement and control of the colour of lamps is also important.

Good colour control in the packaging industries has become more necessary since the advent of the serve-yourself supermarket. One is confronted on the shelves with numbers of similar coloured packets at close range. Slight differences of colour if they existed would easily be seen, and these could be interpreted wrongly as differences in the product, or in the date of its production.

Other examples are the need to control the colour of pots of paint so that the paint from one pot will match that from another with the same colour name. The colours of coloured signal lights obviously must be standardized and controlled by accurate measurement. A cursory look at the large range of colours exhibited by the red rear lights on motor vehicles will quickly convince one that there is need for a tightening of colour control in this case.

7.2 THE PRINCIPLES OF COLOUR MEASUREMENT

The basic principle of all colour measurement is the fact that a match to most colours can be made by the addition of suitable quantities of each of three primary colours as demonstrated by Thomas Young (section 4.4, p. 67). If we measure the quantity of each primary colour used to produce a match, these values represent a measure of the colour.

There is of course an infinite choice of primary colours. They can be monochromatic lights, filtered white light, or even coloured papers as used by Maxwell. It is true that the use of different primaries will result in a different specification of the colour. What is in fact done is to use an internationally agreed system with standardized primaries, and measurements made using different primaries can be converted by calculation into standard measurements using the standard primaries.

Let us assume that we wish to specify a colour C in terms of three primaries R, G and B produced by placing suitable red, green and blue colour filters in front of each of three projectors (Fig. 7.1). The colour C is projected on the screen and adjacent to it a similar spot is illuminated simultaneously by the three primary colours R, G and B. We now vary the amounts of R, G and B so that the two spots of light match both in

The perception of light and colour

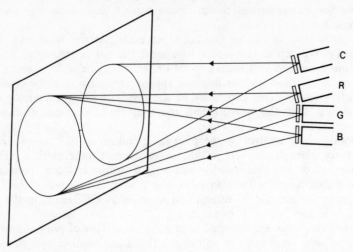

Fig. 7.1 Matching a test colour with an additive mixture of red, green, and blue lights.

colour and in luminosity. If we assume that we have C lumens* of C, and we obtain a match when R lumens of red, G lumens of green and B lumens of blue illuminate the other spot, we can then say that:

$$C(C) \equiv R(R) + G(G) + B(B), \qquad (7.1)$$

where $C(C)$ means C lumens of colour C; $R(R)$ means R lumens of red, etc.; and the sign $\equiv$ means 'matches in colour and luminosity'.

It can happen, as for instance was found by Helmholtz, that if C is a fairly saturated blue-green, no additive mixture of R, G and B can be made to match it. Any mixture will be found to be too white or de-saturated. What is done in this case is to swing the red projector across so that the light from it falls on to the colour spot C, and it will then be found that suitable amounts of G and B will cause the two spots to match (Fig. 7.2). Let these amounts be R_1, G_1 and B_1 respectively. Our equation then becomes:

$$C(C) + R_1(R) \equiv G_1(G) + B_1(B).$$

We can, however, treat these colour equations as ordinary algebraic equations, and we can rewrite this as:

$$C(C) \equiv - R_1(R) + {}_1(G) + B_1(B).$$

* It is usual in these discussions to refer to the luminous flux in lumens which enters the eye. However, we often work in terms of the luminance of a surface, and it would be equally valid to substitute units of luminance such as cd m^{-2} or any other suitable units which are proportional to one another. (See Glossary of Terms, p. 180).

In other words the colour equation for C has a negative quantity on the right-hand side. It was indeed the presence of these negative quantities which first worried Helmholtz, since Thomas Young's three 'nerves' could only signal positive values. However, as we saw in Section 4.8, p. 70, using Wright's metallurgical analogy, these negative values can be explained. If we refer to this analogy, by adding red to our sample, we are in the analogous state to that in which we needed to add some tin to our brass sample in order to match it.

It is convenient to calibrate our colour measuring apparatus by standardizing it with a match on some specified white. In this case, suppose W lumens of white (W) are matched by adding R_w lumens of red to G_w lumens of green and B_w lumens of blue. Our equation for this match with white would then be:

$$W(W) \equiv R_w(R) + {}_w(G) + B_w(B) \tag{7.2}$$

If we took actual figures we would find that a typical match would be where 230 lumens of white were matched by 62 lumens of red, 162 lumens of green and 6 lumens of blue, or:

$$230(W) \equiv 62(R) + 162(G) + 6(B) \tag{7.3}$$

It is fairly typical of colour matches that the amount of blue required is very small compared with the other two primaries when measured in

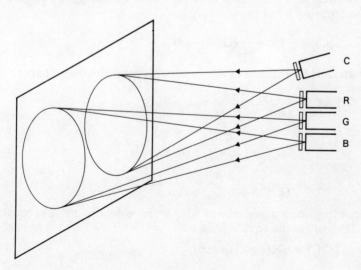

Fig. 7.2 Matching a test colour with an additive mixture of blue and green by adding red to it.

photometric units. A step is taken at this point which is not allowed in normal physical equations. That is to change the units in which each term of the equation is measured. This is done because white seems to be biased towards no colour more than another. If we now make our unit of red 62 lumens, our unit of green 162 lumens and our unit of blue 6 lumens, equation 7.3 can now be rewritten in these new units as:

$$240(W) \equiv 1 \cdot 0(R) + 1 \cdot 0(G) + 1 \cdot 0(B) \tag{7.4}$$

This has the advantage that by inspection the right-hand side of equation 7.4 can be seen to be white, whereas anyone could easily imagine equation 7.3 to represent a fairly saturated yellow green.

Equation 7.4 still represents a given number of lumens of white, but it is convenient at this point to dissociate the colour from its luminance, since a higher or lower luminance of the same colour will have the coefficients of (R), (G) and (B) in the same ratio, but with different values called the tristimulus values. It is the ratios between the values which determines the colour. The colour and luminance are separated by dividing the left-hand side of equation 7.4 by 240, and the right-hand side by the sum of the coefficients, i.e. by 3.0. This gives:

$$1 \cdot 0(W) \equiv 0 \cdot 333(R) + 0 \cdot 333(G) + 0 \cdot 333(B) \tag{7.5}$$

This is known as a unit trichromatic equation, and the unit of (W) is known as 1 Trichromatic Unit or 1T Unit.

Returning now to the algebra of equations 7.1 and 7.2, we can express 7.1 in terms of 7.2, thus:

$$C(C) \equiv \frac{R}{R_w}(R) + \frac{G}{G_w}(G) + \frac{B}{B_w}(B)$$

and dividing by the sum of the coefficients on each side we have:

$$1 \cdot 0(C) \equiv \frac{R/R_w}{(R/R_w) + (G/G_w) + (B/B_w)}(R)$$

$$+ \frac{G/G_w}{(R/R_w) + (G/G_w) + (B/B_w)}(G)$$

$$+ \frac{B/B_w}{(R/R_w) + (G/G_w) + B/B_w)}(B) \tag{7.6}$$

which expresses in a general way what we have done to arrive at equation 7·5. We can rewrite equation 7.6 as:

$$1 \cdot 0(C) \equiv r(R) + g(G) + b(B), \tag{7.7}$$

where $r + g + b = 1$.

r, g and b are known as the chromaticity co-ordinates. The fact that

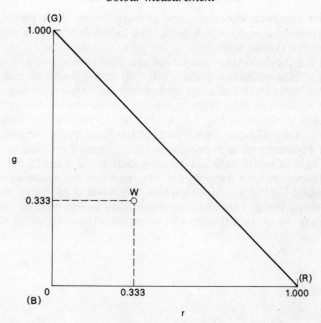

Fig. 7.3 An r, g chromaticity diagram.

these sum to unity means that only two need to be specified. Thus if r and g are specified, $b = 1 - (r+g)$. This fact enables us to use a more convenient graphical representation of the colour than Maxwell's equilateral triangle (Fig. 4.8). In fact we can use ordinary orthogonal Cartesian co-ordinates for our colour maps, and plot only two values say r and g (Fig. 7.3). Pure primary red plots at $r = 1 \cdot 0$, $g = 0 \cdot 0$, pure primary green at $r = 0$, $g = 1 \cdot 0$, and pure primary blue at $r = g = 0$. White plots at $r = g = 0 \cdot 333$. This colour map is called a chromaticity diagram. The centre of gravity law enunciated by Maxwell (see section 4.10, p. 72) also applies to this diagram, and additive mixtures of two colours lie on the straight line joining them. The line (R) (G) is the locus of points for which $r+g = 1$, or $b = 0$. This is the locus of saturated greenish yellow, yellow and orange colours. Similarly the line (B) (G) where $r = 0$, is the locus of the blue-green colours, and the line (B) (R) where $g = 0$ is the locus of the purple colours.

7.3 THE USE OF MONOCHROMATIC PRIMARIES AND TEST COLOURS

Although it is not the easiest way practically, the easiest theoretical way of setting up a system of colour measurement is by using monochromatic primaries and monochromatic test colours. These latter represent the most saturated colours which exist, and therefore they give the boundaries of our colour map.

If we use the following monochromatic lights respectively as our red, green and blue primaries namely 650, 530 and 460 nm, we can match every monochromatic radiation throughout the visible spectrum with an additive mixture of these three primaries. The results can be represented graphically as shown in Fig. 7.4, where the intersection of a vertical line at a certain wavelength λ with each of the three curves represents the number of lumens of each primary needed to match a radiant power of 1 W of light of wavelength λ. If we now change our units by dividing all the ordinates on each curve by the area under it, we obtain equality of co-ordinates for the match with white. The white used in this case is an equal-energy white. This has constant radiant energy in each elemental wavelength band throughout the spectrum. This change of units is

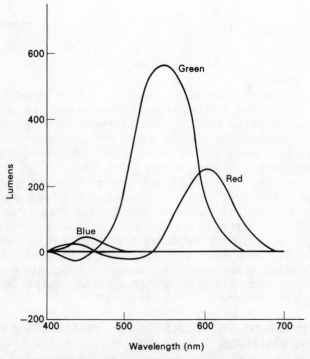

Fig. 7.4 The number of lumens of three monochromatic primary colours needed to match monochromatic lights throughout the spectrum.

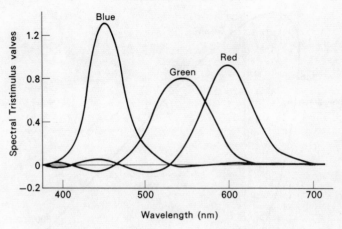

Fig. 7.5 Red, green, and blue spectral tristimulus values.

equivalent to the derivation of equation 7.4 and we now obtain the curves shown in Fig. 7.5, the ordinates of which are known as the spectral tristimulus values. A vertical line drawn at each value of wavelength λ gives the relative values of the chromaticities of λ where it intersects the three curves. These values divided by their sum will give the actual chromaticity co-ordinates. These chromaticity co-ordinates can be plotted on our chromaticity diagram, and when they are joined up they give a horse-shoe-shaped curve which is called the spectrum locus (Fig. 7.6). If we join the red and blue ends we obtain the line along which saturated purples lie, and the enclosed space represents the area in which the whole gamut of colours at a certain luminance will plot. No real colours will plot outside.

7.4 THE CIE SYSTEM

In 1931 an international standardizing body known as the Commission Internationale de l'Éclairage (the International Lighting Commission; referred to as the CIE) recommended a system of colorimetry based on the work of Wright and Guild which has been used with little modification ever since. There are a number of advantages of the system; the most important is that no negative terms appear in the colour equations. This reduces errors when writing down the equations. This can only be done by choosing primaries which are not real colours, which is often difficult to accept by new-comers to the subject. However, it becomes easier to understand if it is realized that although these are purely imaginary as colours, they exist as real points on a colour diagram. Measurements are of necessity made with real primaries, and algebraic

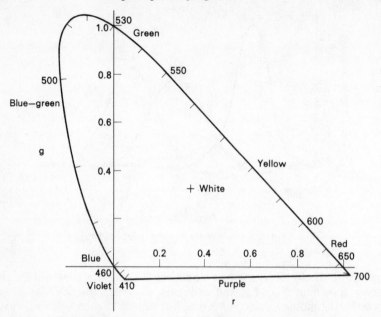

Fig. 7.6 The spectrum locus on an r, g chromaticity diagram. Numbers on the boundary
locus are wavelengths in nm for monochromatic light.

juggling is used to transform these measurements into terms of the unreal
primaries.

If we now inspect Fig. 7.6, and take any three spectral colours as
primaries and join them by three straight lines, we obtain a triangle.
Due to the curvature of the spectrum locus, some colours will always fall
outside this triangle, wherever we choose our primaries. Colours falling
outside this triangular area will need some negative values in their colour
equations. The CIE, therefore, chose unreal primaries in the positions
X, Y and Z so that the whole of the area between the spectrum locus
and the purple line lies within the XYZ triangle. As we stated above,
although these primaries are theoretical and cannot be realized in
practice, the transformation to them of measurements made using real
primaries is purely a matter of linear algebra. So although colour
measurement must be done with real primaries, it means very little effort
to convert to the unreal primaries by calculation or by using tables or
graphical means. It has the added advantage that measurements can be
made in any chosen primaries, but when they are converted to standard
primaries, measurements made in different laboratories with different
primaries can be compared, whereas otherwise they could not. The

choice of the CIE primaries leads to the standard CIE chromaticity diagram as shown in Fig. 7.7. In fact the whole CIE system of colour measurement can be represented graphically by a set of spectral tristimulus values similar to those shown for real primaries in Fig. 7.5. These are shown in Fig. 7.8. A vertical line drawn at any wavelength λ, where it intersects the three curves gives the relative values of the X, Y, and Z primaries needed to match that particular wavelength λ. These are called the CIE spectral tristimulus values $\bar{x}$, $\bar{y}$ and $\bar{z}$, and each divided by their sum give the CIE chromaticity co-ordinates for spectral colours which are referred to by the letters x, y and z. Maxwell was in fact the first to use these symbols for colour equations, although they referred to his real spectral primaries. In a paper published in 1860 in which he described his colour box, he used the symbols X, Y and Z for the slits which he placed in the red, green, and blue parts of the spectrum. He used x, y and z for the corresponding values in his colour equations.

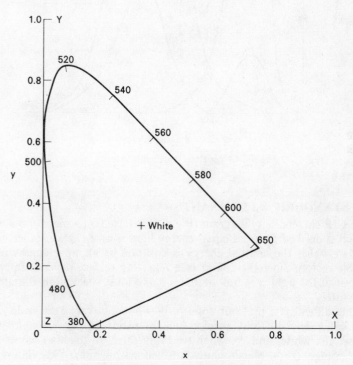

Fig. 7.7 A CIE x, y chromaticity diagram.

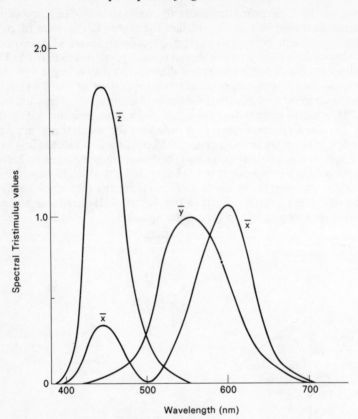

Fig. 7.8 CIE spectral tristimulus values.

7.5 STANDARD ILLUMINANTS

The CIE chromaticity diagram (Fig. 7.7) is based on a match on a white which is derived from an equal-energy light source (S_E), in other words one in which the emitted energy is constant at all wavelengths in the visible region. Since the colour of a reflecting surface depends upon the colour of the light which illuminates it, we must specify our illuminants accurately.

One difficulty is that our concept of white can vary very widely. We can call many different articles such as sheets of paper, paint, wool, snow, etc. white, and yet when we bring them all together we see that they range from bluish-whites to yellowish-whites. One interesting experiment is to make two holes in a dark window shutter over which opal screens are placed. Through one daylight is allowed to enter, but behind the other a tungsten filament lamp is placed. Seen in juxta-position neither looks white; the daylight looks blue and the tungsten light yellow. The quality of 'white' varies a great deal throughout the day appearing reddish in the early morning and late evening and bluish during the day. The colour also depends upon whether direct sunlight is included since this is yellow compared with blue sky light. For these reasons, the CIE specify several standard white illuminants.

Standard illuminant A (S_A) is typical of the light provided by tungsten filament lamps. It is a gas-filled tungsten lamp the filament of which operates at a colour temperature* of 2856 K. Two other standard sources illuminants B and C (S_B & S_C) respectively, represent sunlight and daylight from an overcast sky. They are realized in practice by using source S_A with liquid colour filters of specified composition and thickness.

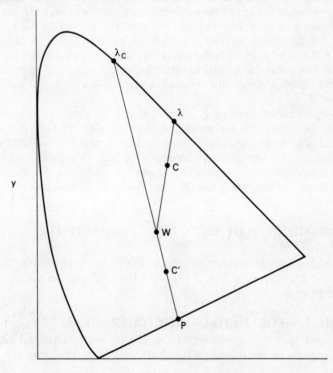

Fig. 7.9 Dominant wavelength and purity.

* See Glossary of Terms, p. 180.

However, sources S_B and S_C are somewhat deficient in blue, and recently for this reason other daylight sources have been specified. In particular source D_{65} is typical of average daylight which is of similar chromaticity to a source of colour temperature 6500 K.

7.6 DOMINANT WAVELENGTH AND PURITY

Unless one is very familiar with the chromaticity diagrams, it is not easy to visualize a colour from its x, y specification, whereas a fair number of people are familiar with the colour of monochromatic lights of various wavelengths. For this reason one often refers to the position of a colour in the chromaticity diagram by its dominant wavelength (λ) and its purity (p). Thus in Fig. 7.9, if we wish to specify the position of a colour at C, we can draw a line WC from a specified white point (W) and produce it to meet the spectrum locus at λ. Since λ, C and W are in the same straight line, the colour C can obviously be matched by a mixture of suitable amounts of the monochromatic radiation λ and the white W. The amounts of each depend upon the position of C along the line. λ is called the dominant wavelength, and the ratio $WC/W\lambda$ is called the purity p. Thus a purity of zero means the colour is white, and a purity of unity, is the value for a fully saturated spectral colour. The purity (an objective measure) is thus related to the saturation (a subjective measure).*

A purple colour such as C' (Fig. 7.9) has no dominant wavelength because the line WC' when produced does not meet the spectrum locus, but meets the purple boundary line at P. In this case we produce the line $C'W$ back to meet the spectrum locus at λ_c. This is called the complementary† wavelength to C', in other words, λ_c added to C' in suitable proportions will make a white W. The purity in this case is defined as WC'/WP.

7.7 COMMON COLOURS IN THE CHROMATICITY DIAGRAM

It is useful to see where common colours fall in the chromaticity diagram, and some are shown in Fig. 7.10. Note the shift on the diagram as a tomato ripens.

7.8 THE UNIFORM CHROMATICITY DIAGRAM

It is found that the sensitivity of the eye to colour changes varies in different parts of the chromaticity chart, and in particular in the green

---

* See Glossary of Terms, p. 180.
† Any two colours which added together produce white are known as complementary colours.

region quite large changes in position are not very obvious visually. To correct this, the chart can be altered in shape so that equal distances on it more nearly represent equal perceptual steps. This is done by algebraically transforming to a different system of co-ordinates namely *u*, *v* and *w* which are related to *x*, *y* and *z* by the following relationships. These were suggested by MacAdam in 1937 and were adopted by the CIE in 1960:

$$u = \frac{4x}{-2x + 12y + 3}$$

$$v = \frac{6y}{-2x + 12y + 3}$$

$$w = \frac{3y - 3x + 1.5}{6y - x + 1.5}$$

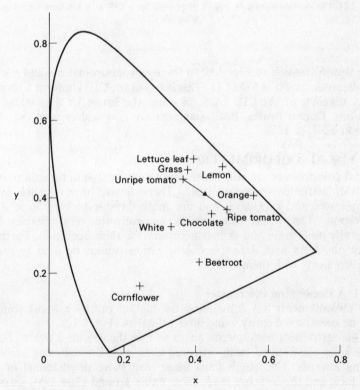

Fig. 7.10 Some common colours plotted on a CIE *x*, *y* chromaticity diagram.

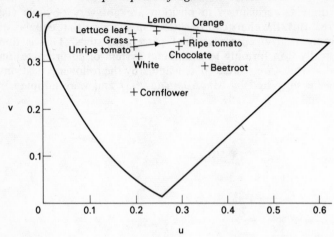

Fig. 7.11 The colours shown in Fig. 7.10 plotted on a CIE u, v Uniform Chromaticity diagram.

The transformation of Fig. 7.10 to these new co-ordinates u and v gives the diagram shown in Fig. 7.11. This is called the CIE Uniform Chromaticity diagram or the CIE UCS diagram; the letters UCS standing for Uniform Colour Scales. Both diagrams are now widely used (see also section 6.5.1, p. 105).

7.9 VISUAL COLORIMETERS

Visual colorimeters are still widely used even though automatic photo-electric instruments are available. These latter, however, are more complicated and expensive, and are more difficult to keep in reliable operation. The visual colorimeters however are very reliable, but naturally more time and skill is necessary for their operation. Furthermore observers with defective colour vision cannot be used to make measurements with them.

7.9.1 A simple filter colorimeter

For measurements not demanding the highest precision a colorimeter can be constructed easily using filter primaries (Fig. 7.12).

This instrument uses mirror boxes to mix the coloured lights. These have sides constructed from mirrored glass with the reflecting surfaces facing inwards. The length must be at least twice the diagonal of the end face, and the ends are made from either ground glass, opal glass or opal plastic. Light entering is well mixed with little loss, by diffusion at

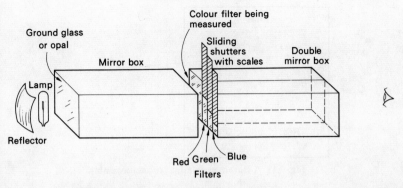

Fig. 7.12 A simple filter colorimeter.

the end faces, and by multiple reflections from the mirrored walls. Light from the lamp is fed into a large mirror box the far end of which becomes evenly illuminated. Half the light then passes through the colour filter to be measured and then into one half of a double mirror box. The other half is covered by the three primary red, green, and blue filters* which are all of the same size and each can be covered by a sliding shutter. The colour seen from the other end is an additive mixture of the three primaries, and the shutters are adjusted until the two ends of the double mirror box match in luminosity and colour. The sliding shutters have linear scales attached to them which indicate the length, and thus the area of the exposed part of each filter. The amount of red, green, and blue in any mixture can thus easily be measured. It may be necessary to have a sliding shutter over the test filter to obtain a luminosity match. Such a simple colorimeter can be adapted for measuring the colour of reflecting surfaces and of after-images (see section 9.4, p. 149).

If it is desired to have only two controls rather than three, a design due to Burnham can be used. This uses a mirror box in the same way, but the filter assembly is different and is made as shown in Fig. 7.13. A fixed rectangular mask which is shown dotted is set with its opening coincident with one end of the mirror box, and the whole filter assembly is moved in two directions. An x-displacement varies the proprotions of red and green, whilst keeping the blue component constant. Since red and green added together give yellow, a y-displacement varies the proportions of blue and yellow, whilst keeping the ratio of red and green constant. This could be thought of as an opponent colorimeter since

* See Appendix for details, p. 183.

The perception of light and colour

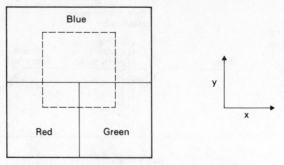

Fig. 7.13 A Burnham colorimeter filter assembly.

x and y movements give separate responses in the red-green and yellow-blue opponent visual mechanisms. Scales attached to the two movements can be used to read the appropriate proportions in a colour mixture.

7.9.2 The Lovibond Tintometer

A very widely-used and excellent commercial visual colorimeter is the Lovibond Tintometer which works on a subtractive principle. The system originated in a simple set of graded brown glasses used to measure the colour of beer. It has since been greatly elaborated and is now based on sets of magenta, yellow, and cyan coloured glasses which supply the subtractive primaries. These coloured glasses are very precisely made, are exceedingly stable, and will hold their calibration indefinitely. The filter glasses are graded from very slightly coloured to those very deeply coloured. Each filter is numbered, according to its depth of colour, on a special scale which has two advantages. First, the numbers are additive so that if several filters are used together, the sum of the numbers will give the same colour as a single filter with the same number. Secondly, it is arranged so that equal quantities of magenta, yellow, and cyan when superimposed give a neutral when they are viewed against a white background. This system enables a match of both colour and luminosity to be made by the superimposition of suitable magenta, yellow, and cyan filters.

However, it is often more convenient to use only two different types of coloured filters at the same time, and to vary the luminosity by a separate control. This method is used in the modern instruments, and it facilitates the conversion from Lovibond colour units to CIE chromaticity co-ordinates.

In the most recent Tintometer colorimeter light is conducted from an internal lamp to the sample along a flexible bundle of very thin glass

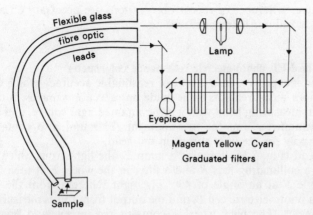

Fig. 7.14 The optical arrangement of the Lovibond Tintometer flexible optic colorimeter.

fibres (a fibre optic) by total internal reflection. The light reflected from the sample is then conducted back along a similar bundle to the eyepiece, in which is seen a semi-circular field illuminated by the reflected coloured light. The colour of the other half is varied by rotating the wheels containing the graduated magenta, yellow, and cyan filters until a match is achieved (Fig. 7.14).

Although this is a visual instrument on which measurements take a little time, need some skill, and cannot be made by an observer with defective colour vision, nevertheless the instrument is relatively inexpensive, and is extremely convenient, stable and reliable.

7.10 PHOTO-ELECTRIC COLORIMETERS

We can nowadays replace the observing eye by a photo-electric cell providing the latter is made to respond to the different parts of the spectrum in an exactly similar way to the average human eye. These devices relieve the observer of tedium and fatigue, and are essential if one needs automatic control of colour, as for example when monitoring a product such as jam or paint in a continuous industrial process. What is not always appreciated is that since we are using photo-electric cells as substitutes for the eye, their performance must be calibrated against visual observations. Measurements from such devices cannot therefore be more accurate than from visual instruments. The measurements with photo-electric instruments have, in fact, higher precision. In other words their scatter about the mean is smaller than for visual measurements. The means, however, are likely to be equally near to the correct result, which is what is meant by equal accuracy. Nevertheless to obtain a good mean value, a number of repeat observations are necessary on a visual instrument and this is time consuming.

7.10.1 The EEL photo-electric reflectance colorimeter

This is an inexpensive instrument of reasonable accuracy which enables rapid colour measurements to be made on reflection samples. It consists of a small measuring head with a self-contained light source, and a separate galvanometer unit for measurement. This unit also contains the power supply for the light source in the head.

The head (Fig. 7.15) contains a lamp L, the light from which passes through a collimating lens K and a filter in the wheel F, it then falls on the sample S at an angle of 45°. The light reflected from the sample falls on a photo-electric cell P, and the output from this is measured on a galvanometer. The filter wheel F contains red, green, and blue filters corresponding to the CIE X, Y and Z colour functions. The CIE chromaticity co-ordinates of the sample can be found quickly by three simple reflectance measurements. The X-filter is first rotated into position and the head is placed over the standard white surface. This is a freshly-cleaned surface of a magnesium carbonate block. The sensitivity control on the galvanometer is adjusted until a reading of 100 is obtained. The head is then placed on the coloured surface to be measured, and the galvanometer reading R_x is noted. This process is repeated with the Y- and Z-filters, the galvanometer reading on the standard white surface being made 90·8 and 32·3 respectively. The corresponding sample

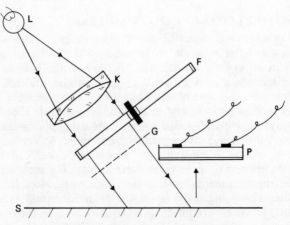

Fig. 7.15 The EEL reflectance colorimeter.

readings R_y and R_z are noted. The CIE chromaticity co-ordinates are given by:

$$x = \frac{R_x}{\sum R}, \quad y = \frac{R_y}{\sum R}, \quad \text{and} \quad z = \frac{R_z}{\sum R},$$

where $\sum R = R_x + R_y + R_z$.

The explanation of this procedure is that the tungsten lamp in the head operates at a colour temperature of 2856 K, and this is therefore a CIE standard illuminant 'A' source. The chromaticity co-ordinates of the source are $x = 0.448$, $y = 0.407$ and $z = 0.145$. The three filters X, Y and Z together with the spectral qualities of the photo-cell give approximately the $\bar{x}$, $\bar{y}$ and $\bar{z}$ colour functions of the CIE system. Since the standard white magnesium carbonate is approximately neutral in colour, that is it reflects equally in all regions of the spectrum, its colour co-ordinates must equal those of the illuminant, i.e. standard illuminant 'A'. This is ensured by setting the galvanometer readings for the white block in the ratio of $100:90.8:32.3$, which is the ratio of the illuminant 'A' chromaticity co-ordinates $0.448:0.407:0.145$.

Since the readings obtained with the filters are always proportional to the three tristimulus values, the above procedure ensures that the correct chromaticities are obtained for all other colours. The reflectance of the sample is measured using only the Y-filter, and initially setting the galvanometer reading for the white block to 97.3 (97.3 is the percentage reflectance of the white block in standard illuminant 'A' light). The head is then transferred to the test sample, and the galvanometer reading gives its percentage reflectance directly. This procedure is followed because the luminance of a sample in the CIE system is represented wholly by the $\bar{y}$ colour function. The $\bar{x}$ and $\bar{z}$ functions indicate only the colour content of the sample and not its luminance.

A number of more elaborate and more accurate commercial instruments are available using similar principles of colour filters and photoelectric cells. Although apparently these instruments are simple, they are not easy to make because the production of colour filters with the correct spectral absorptions presents considerable difficulties.

A number of commercial instruments only measure small differences of colour from standard colour samples, since this is the type of measurement which is often required in mass industrial processes. Such instruments are called colour difference meters. These are more accurate than absolute colorimeters. This arises from the fact that the standard colour can be chosen to be quite close to the test colour, in other words small colour differences only need to be measured. Even as large an instrumental error as ± 10 per cent represents only a very small error in a

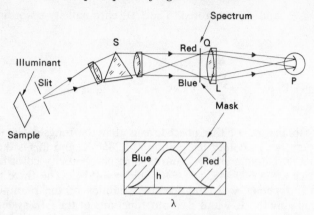

Fig. 7.16 A spectrum mask colorimeter.

small colour difference, whereas it would represent a large error in an absolute colour measurement.

7.10.2 The spectrum mask colorimeter

An ingenious method has been used to overcome the difficulty of making these accurate colour filters. It is based on the fact that if the light from a sample is spread into a spectrum, the light can be attenuated at will by placing a mask (or templet) in the plane of the spectrum. Such a system is shown in Fig. 7.16. The light from the sample is passed through a slit and dispersed into a spectrum at Q by the spectrometer system S. The dispersed light is recombined by a lens L and falls on the photo-electric cell P. In the plane of the spectrum a metal mask is placed from which a section is cut out. The transmission of the system at any wavelength λ is determined by the height h of the appropriate cut-out section of the mask. This value of h can be calculated at each wavelength from the photo-cell response when a slit is allowed to traverse the spectrum, or if a slit of adjustable height is available, this can be varied to give the correct response at each wavelength, and the height noted.

In theory it should therefore be easy to cut out three separate masks so that the overall response of the system and the photo-cell corresponds to the three CIE $\bar{x}$, $\bar{y}$ and $\bar{z}$ functions. In practice, it is not so easily done since the spectrum is usually very small, and this fact necessitates high accuracy in the cutting and locating of the masks. However, it is not too difficult to construct such an instrument providing one does not aspire to very great accuracy. It would be a very interesting and informative project for a more advanced student.

8
The perception of colour

8.1 THE LIMITATIONS OF COLOUR MEASUREMENT

As we stated in Chapter 7 there are great needs in industry for colour measurement. The usual aim is to predict whether two colours are the same so that when they are placed adjacent to one another under standard lighting and viewing conditions their colour will be seen to match. Although we can predict roughly from the CIE chromaticity co-ordinates what hue and saturation a colour will appear, we cannot do this precisely. In certain circumstances our prediction may be a long way out. In other words a colour measurement gives a rather poor indication of how a colour appears to us, that is it gives a poor prediction of our actual perception of the colour.

8.2 COLOUR APPEARANCE

The appearance of colour depends upon a number of factors which we shall discuss in turn. First, it depends upon the luminance level. At very low luminance we have no colour vision, since the rod receptors in our retinae are insensitive to colour. As the luminance is increased the cone receptors come into operation. But of course, as the illumination increases, the receptors give greater signals and the colour difference signals to our brains become stronger. The difference between the colour range and strength of colours of objects on a bright sunny day in the summer, and the dullness of the same objects seen on a grey winter day are well known. It is also common experience that the saturation, i.e. intensity of apparent colour content of red signal lights and neon signs is very much more striking than the smaller range of apparent colour experienced when looking at coloured papers of similar chromaticity. These are always of course of much lower luminance.

Another factor is field size. In small fields the perception of blue disappears. This can be demonstrated by viewing the blue and yellow pattern shown in Plate 7 (*bottom*), and then slowly moving away from it. Beyond a distance of about 1 m the perception of blue will be seen to disappear. This occurs when the object size subtends an angle of about

131

20 min of arc at the eye. This effect is known as small field tritanopia (see section 10.3.1, p. 156).

The next factor is the luminance of the surround. If Plate 8 (*top*) is viewed from a normal distance it will be seen that the appearance of the pairs of identical patches of colour are different when they are surrounded by a bright surround from when they are surrounded by a dark background. The black surround increases the luminosity of the colours (see section 3.9.2, p. 52) but at the same time seems to remove colour, making the patches appear less saturated. This is undoubtedly due to the stimulation of receptors on the parts of the retina which surround the fixation point and which are feeding information across to the fovea. The electron microscope has revealed an enormous number of horizontal connections in the neuron layers of the retina, hence there is a strong physiological support for this type of interaction.

The effects of a light and a dark surround can be demonstrated very well using colour filters fixed to the opal screen of a brightly illuminated photographic viewing box. It is more striking than with the lower luminances which are all that are possible using reflecting paper surfaces. This effect has practical applications. It is possible to obtain a greater gamut of colour perception if a colour television set is placed in a lighted room (with the brightness, i.e. luminance, controls slightly raised) than when placed in a darkened room. This is the opposite result to that which one instinctively expects.

The next effect is that due to the colour of the surround. In Plate 8 (*bottom*) pairs of the same colour are shown on a yellow and a blue background. The change in the appearance of the colours is striking, and this is also due to the cross connections in the retina feeding colour information to adjacent colour channels. It is known as simultaneous colour contrast. Complicated colour patterns produce the most amazing effects. Plate 9 (*top*) shows a design by S. Harry. There are only three colours used to produce this design, but complicated interactions produce many different colour perceptions. The effect differs with viewing distance. A longer viewing distance results in the reduction of distances between the coloured patches in the retinal image. This causes a greater degree of interaction, and thus a greater modification of the perceived colours. If the diagram is viewed at very close range, these different effects largely disappear, and it is quite obvious that only three basic colours are present.

Another interesting effect is seen when viewing Plate 9 (*bottom*) at a distance. This figure consists of alternate red and blue bars. If seen at such a distance that the bars are not resolved (5 m or more) the coloured area appears purple as one would expect. However, the purple colour will be seen to some extent at even shorter distances, when the separate red and blue bars can be seen. This is due to the fact that the normal person is considerably myopic in blue light due to the uncorrected chromatic aberration of the eye. Thus, whereas red light focuses on or near the retina, blue light focuses an appreciable distance in front. This gives rise to a large blurred blue image on the retina, which overlaps the red image, thus resulting in the perception of a purple.

Another effect is shown in Plate 10 (*top*). The background colour is the same across each strip, but when it is overlaid with a fine black pattern it appears darker, and when it is overlaid with a fine white pattern it appears lighter. This is the opposite effect to that produced by a light and a dark surround. It is known as the Von Bezold spreading effect, because the luminance of the fine pattern seems to spread over into the background.

These effects have practical applications as S. Harry has also shown with his carpet designs. In certain circumstances, for instance, it may be necessary in a complicated design to use *different* colour wools when the *same* colour perception is required. It also means that with ingenious design, many different apparent colours can be produced from a small number of different coloured wools.

Another very striking colour appearance effect is shown by an experiment on coloured shadows which can be carried out very easily. The experimental arrangement is shown in Fig. 8.1. Two projectors are arranged to project two overlapping white areas of light onto a white

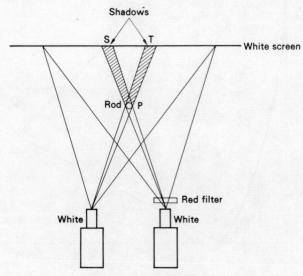

Fig. 8.1 Coloured shadows.

screen. One is coloured red by placing a red filter over the projection lens. An opaque vertical rod is placed in the light beams so that two shadows are cast on the screen. The shadow S formed by the red projector is illuminated only by the white light, whilst the shadow T formed by the white projector is illuminated only by red light. Thus the shadow S should appear white and the shadow T, red. In fact S appears to be a very distinct blue-green and T appears red. This surprising result is due to the fact that most of the screen is illuminated by pink light which is formed by the mixing of the red and white light, and its colour is approximately at the point R on the chromaticity diagram (Fig. 8.2). The eye quickly adapts to this and tends to perceive this as white. The original white of chromaticity W is then on the blue-green side of R and this appears to have a dominant wavelength λ which is blue-green. The shadow T also changes but to a similar red. Of course, the chromaticity of S has not altered but has only appeared to have done so. It is interest-

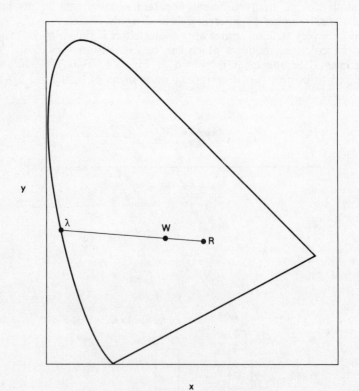

Fig. 8.2 Explanation of coloured shadows.

ing to replace the red filter by a green filter. The shadow *S* now appears red. If a blue filter is used, *S* appears yellow.

This is an elementary version of a famous experiment by Edwin Land, the inventor of *Polaroid*. He showed that a surprising wealth of colours could be perceived by performing Maxwell's colour photography experiment (see section 5.8, p. 87), but by using two projectors instead of three, and furthermore by using only *one* colour filter. For example, if he projected a *red* image of a scene taken through a red filter on top of a *white* image taken through a green filter, then yellows and greens as well as reds could be perceived in the picture. The effects in a complicated scene are even more surprising than with the simple coloured shadows, and are well worth repeating. Bowls of fruit which include oranges and ripe bananas seem to make good subjects to photograph for such experiments.

8.3 SURFACE COLOURS

There are many fascinating perceptual effects which are conveyed to us by the texture and lustre of surfaces. The texture of a surface conveys the information to us which tells us whether we are looking at stone, brick, cement, plaster, paint, cloth, velvet, wood, etc. and this information is in addition to that conveying the perceptions of colour and luminosity. By texture, of course, we mean small and subtle variations of colour and luminance, and the information is such that it can still convey to us what type of surface it is, even when we are looking at a colour photograph or a colour television screen.

Another interesting factor in the perception of surfaces is that of metallic lustre. Metals can have a matt, semi-polished or fully-polished surface, and the colour of the metallic sheen and highlights assist us in distinguishing which type of surface we are looking at. Furthermore, we can tell whether the surface is copper, brass, gold, silver or chromium or stainless steel by the colours of the reflections.

Yet again very subtle perceptual colour effects are produced by translucent materials such as alabaster. Such materials appear to glow from within and there is a *volume* element added to the colour perception. In other words the colour appears to be emanating from within the material and we no longer think we are looking at a surface colour. However, Ralph Evans has shown that when an area of colour is given a diffuse edge, it appears hazy and almost translucent. This is an area of perception about which we know very little at the moment.

8.4 THE HELMHOLTZ–KOHLRAUSCH EFFECT

Helmholtz was the first to notice that the saturation of a colour affected its luminosity, and that some colours appear much brighter at higher saturations even though the luminance was kept constant. The effect was investigated subsequently by Kohlrausch. He showed that if he placed two patches of colour of the same luminance side by side, one of high and the other of low saturation, the former would appear brighter, but that if they were flickered and thus seen alternately, the differential luminosity effect disappeared.

The effect can give rise to odd situations, which can be demonstrated easily. For example saturated red and green patches of light can be projected separately on to a screen, and made of equal luminosity to two further spots of white light. If the red and green are now superimposed, the resulting yellow patch (the mixture of red and green) will appear considerably less bright than the combined white patches. Also as shown by MacAdam, the removal of some of the red component of a mixture of red, green, and blue lights increases the luminosity of the remaining mixture.

These striking experiments are demonstrations of the failure of the law of addivity of luminance (which implies that a higher luminance always results in a higher luminosity), and in certain circumstances has quite serious repercussions for the science of photometry. Normally, however, one is dealing with coloured surfaces of reasonably low saturation, when luminances are additive.

The effect is of importance, however, when dealing with fully saturated lights of high luminance, for example coloured traffic and railway lights, rear lights, and direction indicators on motor vehicles, and electric discharge-tube advertising signs. It is well known that a bulb of lower power can be used in a red motor car rear light than in an amber turn indicator. This is because the effect is more marked in the red than in the yellow.

The Helmholtz–Kohlrausch effect can be investigated by the method of direct estimation. Results of one such experiment are shown in Fig. 8.3 for a white stimulus, S_1, for colours (S_2) of medium saturation, and for colours (S_3) of high saturation. The ordinates are the logarithms of the ratio of the white to the colour luminance for equal luminosity. These ratios are greater than unity (log ratio > 0), which means that the white has to be of higher luminance than the colour to appear of the same brightness. For some subjects these effects are quite large. For example for the blue at high saturation, the log ratio is 1·70 which means that the white has to be 50 times the luminance of the blue stimulus to appear of the same luminosity. The corresponding figure for the red is 31 times. The effect is much less in the yellow, however, but it can still be appreciable and a white has to be four times the luminance of the yellow to appear equal in luminosity for this subject.

The explanation of the effect is probably to be found in the neural organization of the retina. As we explained in section 4.13, p. 74, the

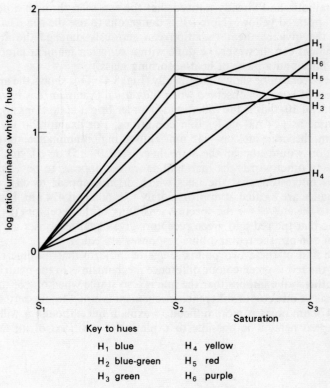

Key to hues

H_1	blue	H_4	yellow
H_2	blue-green	H_5	red
H_3	green	H_6	purple

Fig. 8.3 The Helmholtz–Kohlrausch effect.

luminance information is transmitted to the brain along the non-opponent channels, whereas the colour information is sent as colour difference opponent signals. It seems probable that when saturated colours are observed, the colour difference signals are very strong, and that the perception of luminosity is evolved in the brain from information received not only from the non-opponent channels but also from the colour difference mechanisms.

8.5 THE BEZOLD–BRÜCKE EFFECT

This phenomenon first discovered by Von Bezold in 1873, but later discovered independently by Brücke in 1878, relates to the fact that variation of luminance modifies the perception of hue. Increasing luminance makes red yellower, and violet and blue-green bluer.

Helmholtz had already noticed that the sun seen through a deep red glass appeared yellow. (Note—It is dangerous to use the sun as a source since the invisible heat radiation can seriously damage the eye. The experiment can, however, be done with a tungsten filament lamp, using a red filter and a piece of heat-absorbing glass.)

The effect can be shown graphically (Fig. 8.4). This shows the apparent changes of hue ($\Delta\lambda$) observed when the retinal illuminance is lowered by a factor of 10, that is the radiation (λ) under bright conditions appearing the same as ($\lambda + \Delta\lambda$) under dim conditions. For example in the red at 620 nm, the value of $\Delta\lambda$ is − 15 nm. Thus at high illuminance, the 620 nm radiation would appear the same hue as (620 − 15) or 605 nm at low levels. In other words the high luminance red appears to be yellower.

It is interesting that the hues which do not appear to change with luminance are located at approximately 571, 506 and 474 nm. If we refer back to the curves for the spectral sensitivity of the receptors (Fig. 4.10) we see that the red and green receptors give equal responses at 575 nm, and at 506 nm the red and blue responses are equal.

The first of these two points seems to indicate that the hue is stable when the red *v*. green colour difference mechanism is in the neutral state. The other value implies that the hue is also stable when one of the components of one colour difference mechanism is equal to one of the other. The 474 nm point is more difficult to explain but although it will not be attempted here, it is possible to explain the positions of all three in-

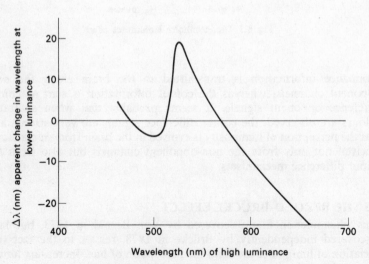

Fig. 8.4 The Bezold–Brücke effect.

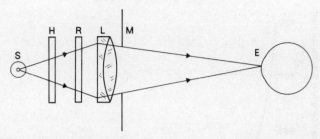

Fig. 8.5 Apparatus to observe hue changes with time for a bright source of light.

variant hues by the assumption of two opponent colour mechanisms in the retina.

A related phenomenon has been reported by Cornsweet. This refers to the change in perceived hue with time as a very bright red light source is viewed. This can be demonstrated very easily either by looking directly at a bright lamp through a red filter together with a heat-absorbing filter, or with the more sophisticated arrangement shown in Fig. 8.5. A low voltage car headlamp S has its filament imaged on the eye pupil at E by a lens L. A diaphragm M is used to limit the field of view to about $5°$. A red filter R (e.g. Cinemoid No. 6) is placed in the beam together with a heat-absorbing filter (Chance HA3) at H. (Note—It cannot be over-emphasized how important it is in both these experiments to use a heat-absorbing filter. This is because filament lamps emit most of their radiant energy as heat, and unless this is removed it can cause damage to the cornea, or even produce a retinal burn, which could very seriously damage the sight of the eye.)

The type of variation of colour experienced is shown in Fig. 8.6 from which it will be seen that after 10 s the lamp appears yellow, then green after 60 s, finally settling down to a yellow colour. To explain this we can assume the red stimulus is of wavelength about 630 nm, and hence the red receptors are very strongly stimulated; the green being less so. At this high intensity, the red receptor photo-pigment would be rapidly bleached which would result in a quick fall off of the red signal to the red-green colour difference mechanism. This would cause it to signal green after a short time. Quite soon a reasonable amount of the green receptor photo-pigment would also become bleached, however, and this would result in a fairly equal response from the red and green receptors, and the red-green colour difference mechanism would assume a neutral state.

There would though still be a comparatively large red + green input to the yellow-blue mechanism which would give the final yellow perception.

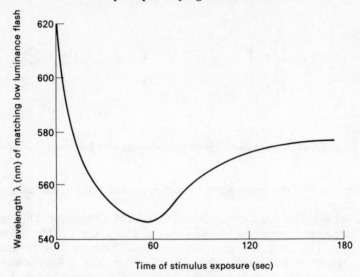

Fig. 8.6 Hue variations with exposure time.

8.6 THE DIRECT ESTIMATION OF HUE AND SATURATION

It is clear that colour measurement, although it is of great value in a large number of industrial situations, is quite unable to record a large number of these perceptual effects which have been outlined in this chapter. The need to assess the veracity of the reproduction of colours by colour television and colour photography has led to recent work on a new method by Rowe. He has found by presenting observers with coloured fields with a standard white surrounding area, that after some training, they were able to estimate directly both the hue and the saturation of the colours seen with quite reasonable reproducibility. This is similar to the direct estimation of luminosity which was described in section 3.11, p. 55.

The hue was estimated in terms of the four psychological colour primaries namely red, yellow, green, and blue. An observer either said pure red or pure blue as the case may be. If he saw a red purple he might say '50 red, 20 blue' to indicate the exact position between pure red and pure blue. The saturation was estimated independently as a percentage, where 0 represented white, and 100 a fully saturated spectral colour.

The estimates of hue were found to be more precise than the saturation estimates. However, it was possible to 'calibrate' each observer in terms of his estimates, and lines of equal hue and saturation estimates are shown plotted on a CIE UCS chromaticity diagram in Fig. 8.7. When

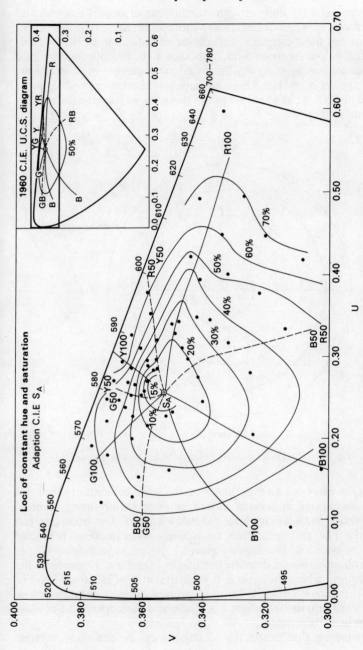

Fig. 8.7 Estimates of hue and saturation on a CIE UCS diagram.

this has been done the observer can estimate hue in an actual scene, and then afterwards on a colour television picture, or on a colour photograph. Whether or not these estimates agree gives an indication of the faithfulness of the colour reproduction, although it is complicated by odd preferences such as wanting sky bluer and grass greener in photographs than they appear in real life. This is perhaps just another example of living in a world of fantasy, in which we prefer things as we imagine them to be, rather than as they are.

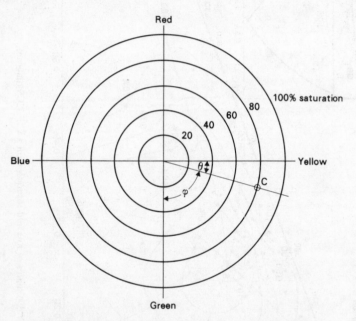

Fig. 8.8 Diagram for plotting estimates of hue and saturation.

It is easy to carry out an experiment on the direct estimation of colours without complicated apparatus. It can be done readily using Munsell papers—whose chromaticity and Value are known. The observers can indicate the hue and saturation on a single diagram (Fig. 8.8). For example the colour C is a slightly greenish yellow of saturation 80 per cent. The observer would describe this hue as '50 yellow, 10 green, 80 per cent saturation' where the ratio $\phi/\theta = 50/10$ (i.e. $\phi = 75°$, $\theta = 15°$). The lightness (luminosity) of these Munsell samples can also be estimated on a scale 0 to 10 or 0 to 100, where 0 is black, and the upper value is white of 100 per cent reflectance.

It is interesting that practically all colours can be described in terms of the four psychological primaries with the notable exception of browns. Browns plot on the chromaticity diagram near the orange-yellows, and they are always of low luminance relative to other colours. We can therefore describe them as low luminance orange-yellows, but they certainly do not look like this. So brown seems to be a complicated perception which arises from viewing low-luminance orange-yellow surfaces against an illuminated background. It is undoubtedly a very interesting phenomenon which results from the way in which the brain interprets the signals sent to it from the retinal colour difference mechanisms when these signals are weak. It is clear that browns can border on the red or yellow, giving reddish or yellowish browns respectively. There is, however, no such thing as a bluish brown, so it appears that it is a phenomenon exhibited at a sufficiently high wavelength for the blue receptor to be inoperative (i.e. above 570 nm).

8.7 THE FUTURE

As Evans has pointed out hue, saturation and lightness are only three of many aspects of human colour perceptions. Depending on the circumstances a colour can have a grey content, it can exhibit great 'brilliance', it can appear metallic, highly polished or matt (with many intermediate variations), it can appear translucent, and also the surface texture can influence the perception. The psychophysical correlates of dominant wavelength, purity and reflectance, are respectively hue, saturation and lightness, but we have as yet no psychophysical correlates for the other percepts. Undoubtedly as time goes on the problem of finding suitable correlates may be solved.

The phenomena and problems we have discussed in this chapter are problems concerning our basic perceptions, which have their origin at many different levels of the visual system. The understanding of the working of the eye and brain is, however, still in its infancy. Until we understand how the brain works in very much greater detail, we shall not begin to unravel these perceptual problems. It may be, of course, that with a brain as a tool we can never get a true insight into how it itself works. One can for instance never directly see one's own eyes, but in this case a mirror gives us a good, but nevertheless reversed image. Is there a similar 'brain mirror' which will allow us to see into our own brains? It is conceivable that in this case the answer is 'no'. Nevertheless man is such a curious animal that research into the brain and perception will and must go on.

9
After-images and subjective colour

9.1 THE PERSISTENCE OF VISION

When we are dealing with everyday events and not astronomical problems, it is reasonable to assume that light travels so fast that its effects are instantaneous. The processes of vision, however, are not instantaneous and it takes about 0·1 s before an observer perceives that a light flash has entered his eye. Likewise too the perception of light persists at the same level for a short time after the light stimulus has ceased. This is known as the persistence of vision.

It is easy to demonstrate this by arranging to view a lamp through a rotating sector disc which has alternate opaque and transparent parts (Fig. 9.1). The sector can be rotated by an electric motor with a speed control. If the light flashes seen through the disc have a reasonably low frequency, say about 10 Hz, a flickering light will be seen. If now the frequency is slowly increased, at a certain value f_c, the flickering sensation will disappear. This occurs when the flashes come sufficiently fast for the persistence of vision to carry over the perception during the time the light is extinguished. This frequency f_c is either called the critical flicker

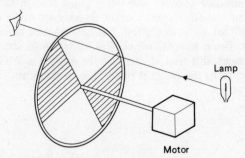

Fig. 9.1 Finding the flicker fusion frequency.

144

frequency (CFF), or a better name is the flicker fusion frequency (FFF). Its value varies linearly with the logarithm of the luminance (L) to which the eye is adapted. It varies from about 10 Hz at low luminance to about 55 Hz at high luminance. Mathematically we can write:

$$f_c = a \log_{10}L + b,$$

where L is expressed in cd m^{-2}. This is known as the Ferry–Porter law, and is a consequence of the electrophysiology of the retina. As we explained in section 3.6, p. 46, luminosity (brightness) is coded as a spike potential frequency output from the retinal ganglion cells, and this frequency is approximately proportional to the logarithm of the lumin-ance. Obviously, if the flashes of light arrive at the eye at a greater frequency than the spike potentials, the latter cannot signal the flash frequency to the brain, in which case fusion results. However, at high luminance the spike potential frequency is higher, and so therefore is the flicker fusion frequency. This short-term persistence phenomenon is exploited in a number of ways. For example vapour discharge lamps and tubular fluorescent lamps flicker at 100 Hz when run from a 50 Hz alternating current supply. (Note—Reversal of the current does not change the sign of the light output, which can only be positive or zero, hence there are 100 light pulses every second on a 50 Hz supply.) This is well above the flicker fusion frequency, and hence in ordinary circum-stances these lamps appear to have a steady brightness. Again a cinema projector projects 24 different pictures (frames) each second. Each picture is exposed twice and thus there are 48 pictures projected on the screen each second with dark intervals in between. The luminance is sufficiently low that the result is the impression of a steady picture, and most people are amazed to learn that the screen is in fact dark for *half* the time.

Television is another example of the use of this phenomenon. In fact at any instant of time only a single small spot of light illuminates the screen. This is the so-called 'flying spot', since it traces out a picture very rapidly by drawing 313 adjacent lines on the screen in 0·02 second (linked to the mains frequency of 50 Hz), in fact it first draws the odd numbered lines 1, 3 5, etc. in the picture. It then very rapidly goes back to the beginning and then draws another picture of 312 lines in 0·02 second which interlaces the first, that is it draws the even numbered lines in the picture: 2, 4, 6, etc. The result is a reasonably steady picture, although some flicker is occasionally apparent, especially if the screen is of high luminance and is viewed by the peripheral retina, where the flicker fusion frequency may be higher. The 60 Hz mains frequency used in some countries is sufficiently high to remove the residual flicker.

These are interesting examples of where a knowledge of the charac-teristics of the visual process has resulted in useful widespread applications. Clearly without this in-built persistence of the visual mechanism, the cinema and television could not exist in their present form.

9.2 TYPES OF AFTER-IMAGE

The persistence of vision phenomena can be said to be due to short-period after-effects or after-images which last only for a fraction of a second. However, if the eye is dark-adapted and then stimulated by a single brief bright flash of light (e.g. by briefly viewing a clear car head-lamp bulb), a faint perception of light in complete darkness persists for a longer time, and gradually decays over a time of one or two minutes. This is known as a long period positive after-image. Unlike the persisting image explained in the last section, although one may think it quite bright, this after-image will only appear about as bright as one-millionth of the luminance of the original stimulus. The positive after-image is initially the same colour as the stimulus, but its colour varies with time, and goes through a number of colour changes as will be explained later.

If the primary high-luminance stimulus is of small angular size, however (e.g. a few degrees across) and when extinguished is immediately followed by a large white adapting field of low luminance, a bright positive after-image will first be seen as before. If now the luminance of the large white field is quickly increased, the positive after-image will first fade into its surround, and then will appear as a dark spot when the surround luminance reaches a higher value. This dark spot is called the negative after-image. The negative after-image of a coloured primary stimulus is approximately complementary in colour. Thus, the after-image of a red stimulus is blue-green, and that of a green stimulus is magenta.

The study of these after-images has fascinated a great number of famous people from Aristotle onwards, and Newton, Boyle, Young, Purkinje, Goethe, Fechner, Helmholtz, and Hering have all written about them.

In 1664 Boyle had described experiments which involved looking at the sun through a telescope, and says he had seen an image of the sun as a white object some 9 or 10 years afterwards! (This was an exceedingly dangerous thing to do and could easily have resulted in the loss of the sight of an eye. In the circumstances, it seems to have resulted in a severe retinal burn, which is of course irreversible and permanently damages the retinal receptors.) Locke read Boyle's account and asked Newton's opinion on the matter. His reply in a letter from Cambridge in June 1691 is interesting:

> The observation you mention in Mr Boyle's book of colours, I once made upon myself with the hazard of my eyes. The manner was this: I looked a very little while upon the sun in the looking glass with my right eye, and then turned my eyes into a dark corner of

my chamber, and winked, to observe the impression made, and the circles of colours which encompassed it, and how they decayed by degrees, and at last vanished. This I repeated a second and a third time. At the third time, when the phantasm of light and colours about it were almost vanished, intending my fancy upon them to see their last appearance, I found to my amazement, that they began to return, and by little and little to become as lively and vivid as when I had newly looked upon the sun.

He goes on to describe how apparently he could transfer the after-image to the other eye by mental effort, although whether this is actually what happened is not clear from his writing. It is probable, however, that he also had produced a mild burn on his retina, because he goes on to say:

I had brought my eyes to such a pass that I could not look upon no bright object with either eye, but I saw the sun before me, so that I durst neither write nor read, but to recover the use of my eyes, shut myself up in my chamber made dark, for three days together, and used all means to divert my imagination from the sun.

He confesses that the problem is 'too hard a knot for me to untie'. We must confess too that nearly three hundred years later we are still puzzling over after-images.

Negative after-images are easily demonstrated by viewing the filament of a lamp with a clear envelope for 2 seconds, after dark-adapting for a minute or two. The positive after-images can be seen projected on to a dark projection screen. If a slide projector is now used to flood the screen with white light, negative after-images will be seen. The positive after-image can be proved to be still there by switching off the white projector. Colour filters, e.g. a green, can be placed over the lamp. The striking alternation of the green positive and magenta negative after-images can easily be seen, and it is interesting that the positive after-image must obviously still be there when the negative after-image is seen.

9.3 NEGATIVE AFTER-IMAGES

The fact that the negative after-image predominates in the continued presence of the positive after-image, means that the former is of considerably higher brightness.

These complementary negative after-images are fairly easily explained by assuming that they are due to the fatigue of small areas of the retina. For instance, if the primary stimulus is green, the green cone receptors will have more pigment bleached than the red or blue cones. They will therefore be less sensitive to green light. Thus, when it is followed by a larger white stimulus of lower luminance, that part of the retina previously exposed to the green stimulus will perceive white with less green than normal. White minus green is perceived as magenta, which is the colour of the after-image. The fact that it is seen surrounded by a larger white area viewed by cones which have not been differentially bleached, enhances the colour perception.

A striking demonstration of negative after-images is Bidwell's disc (Fig. 9.2). This is easily constructed from white card and can be rotated by an electric motor with a speed control. When rotated rapidly in the direction shown, a red lamp will appear a brilliant green, provided the illumination of the disc is suitably adjusted. What is seen is a complementary negative after-image of the red lamp projected against the white half of the disc. The black area gives a brief period of recovery before the next brief red stimulus is exposed. Since the cut-away part of the disc is only a small fraction of the white area, the primary stimulus is of short duration compared with the duration of the after-image. The extremely interesting fact is that in these circumstances the bright primary stimulus is not perceived, and the perception is entirely of the negative after-image.

It is interesting too that an after-image is stabilized on the retina. Normally when we rotate our eyes but not our head, the external world appears stationary. The movement of the image across the retina is compensated for perceptually by the brain by information on the eye movements fed in by the muscle mechanisms. This results in an odd paradox. When the eye is rotated when an after-image is present, the after-image is in a fixed position on the retina, and thus when the normal image movement is perceptually annulled, the after-image is given apparent movement relative to the external field.

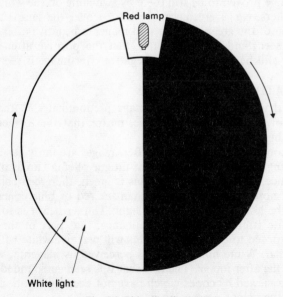

Fig. 9.2 Bidwell's disc.

Another interesting demonstration is that of Emmert's law. This states that the apparent size of a negative after-image is linearly proportional to the distance of the surface upon which it is seen projected. This can be demonstrated by forming an after-image of a clear lamp filament and then viewing a sheet of white paper at a distance of about 25 cm. If a white wall is then viewed at a distance of 3 m the negative after-image will appear much larger. This arises from the fact that the after-image is of constant angular size, and is therefore perceived as the size corresponding to the projection of this angle on to the surface viewed.

The law in fact breaks down beyond a distance of about 200 m, and if the after-image is projected on the clear sky, it does not appear of infinite size. In fact it appears of almost constant size to most people when projected beyond about 200 m. The reason for this is not clear, but it is in fact the same problem as why the sun or moon (which both subtend a constant angle of approximately $\frac{1}{2}°$ at the eye) do not appear infinitely large.

9.4 POSITIVE AFTER-IMAGES

Positive after-images are easily demonstrated by first dark-adapting for 5 to 15 min to allow previous images to subside, and then viewing a very bright source of light for 1 or 2 s. The positive after-image is then viewed in complete darkness.

It is worthwhile setting up a slightly more elaborate light stimulator, one difficulty being with the simplest arrangement is that it is very difficult when in the dark to know where to look for the lamp so that the after-image is produced on the fovea. The arrangement is shown in Fig. 9.3. The dim red light source *S* illuminates a white annulus painted round the diaphragm aperture just sufficiently to hold the fixation before

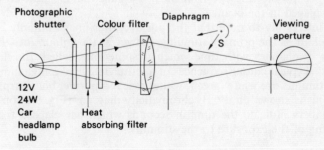

Fig. 9.3 Apparatus for producing a positive after-image.

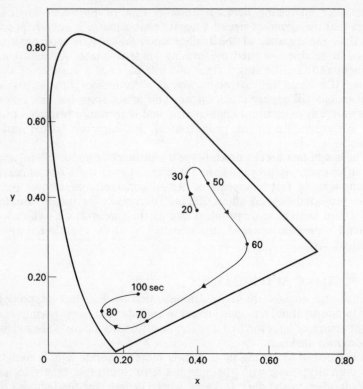

Fig. 9.4 The variations of colour of an after-image of a white light plotted on a CIE chromaticity diagram.

the stimulus is exposed. 'Cinemoid' colour filters* can be inserted as shown. To protect the eye from heat a Chance HA3* heat-absorbing filter is placed in the beam.

It is interesting to note that the positive after-image initially has the same colour as the primary stimulus. However, the colour goes through a series of changes with time which is sometimes called 'the flight of colours'. Typical colour sequences of an after-image of a high luminance white stimulus are white, green, white, red, red-purple, blue-purple, and blue. This is shown on a CIE chromaticity diagram (Fig. 9.4) on which the numbers indicate the time in seconds which has elapsed from the beginning of the exposure to the stimulus.

* See Appendix, p. 183, for details.

It is an interesting fact that whatever the colour of the initial stimulus, the after-image finally appears blue before it fades completely.

The luminosity of the decaying after-image can be measured by a binocular matching technique, similar to that originally suggested by Wright of Imperial College. In this method the after-image is formed in one eye, and its luminosity is matched by a similar adjustable comparison light patch presented to the other eye. These are slightly displaced in a vertical direction so that the two patches can be seen binocularly without fusion. By this procedure the after-image is not affected by the measuring stimulus.

Figure 9.5 shows the retinal illuminance in trolands of the comparison patch which matches the luminosity of the after-image as it decays. The curve shows the effect of increasing the angular diameter of the stimulus

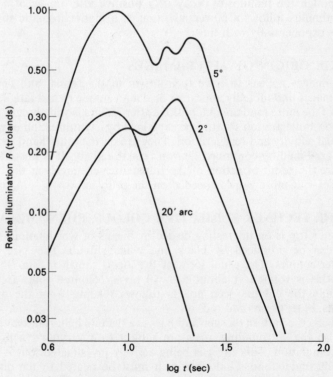

Fig. 9.5 The variation of matching retinal illuminance of after-images formed by three white light stimuli of different angular diameters.

from 20′ of arc to 5°. The larger stimulus gives brighter and more persistent after-images. Also the inclusion of rod receptors in the larger area gives a double peak to the curves. The decay part of the curves for white light stimuli follow the law:

$$R = Ct^{-3},$$

where R is the retinal illuminance in the comparison eye needed to binocularly balance the after-image luminosity, t is the time which has elapsed since the beginning of the stimulus, and C is a constant. It should be noted that the maximum luminosity of the positive after-image is matched by a luminance of only about 0.5×10^{-6} that of the original stimulus. This very low luminosity is one reason why positive after-images are usually not seen in normal vision. Another reason is that even when driving a car at night and looking into oncoming headlamps, after-images are not unduly troublesome, since there appears to be some in-built perceptual mechanism for ignoring them.

Although the luminosity decay of a positive after-image of a white light stimulus follows a power law, that of a monochromatic stimulus decays exponentially with time.

9.5 THE ORIGIN OF AFTER-IMAGES

After-images appear to have their origin in the retina, and negative after-images undoubtedly appear to be due to simple retinal fatigue. The cause of the more fundamental positive after-images is still obscure. Their time-constants are too short to be explained by a continuation of photo-chemical activity and regeneration. They are on the other hand too long to be explained by continued neuron retinal activity. It is possible that they are the result of action of the transmitter chemicals at the retinal synapses, but more work is needed on the problem.

9.6 THE FECHNER SUBJECTIVE COLOUR PHENOMENA

Benham's top is an interesting device by means of which colour sensations can be produced by black and white stimuli. They are called Fechner colours. The usual form of the top is shown in Fig. 9.6, and when this is rotated at about 5 to 10 rev/s, coloured rings are seen. Normally the colours seen are as follows reading from the outside inwards, blue, green, and red.

Viewing the pattern on such a disc gives a definite light-time waveform, and also the surrounding areas are subject to a varying synchronous light stimulation. This effect is being actively investigated, but it is not yet fully understood. Undoubtedly it must be related to the different time responses of the colour and colour-difference mechanisms in the retina.

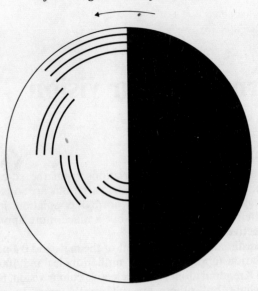

Fig. 9.6 Benham's top.

These phenomena are yet further examples of the very complex structure of the retina and brain giving rise to incredibly fascinating and interesting, but highly complex subjective phenomena.

10
Abnormal colour vision

10.1 INTRODUCTION

Everyone shows small personal variations from the so-called normal colour vision, but with some people the variations are so large that they cannot be regarded as part of the normal population spread. These people are often called colour blind but a much more appropriate term is colour defective.

It has been estimated that 8 per cent of the male and 0·4 per cent of the female population (approximately 2 million males and 100 000 females in the United Kingdom alone) have defective colour vision. Some of these defects are slight but others are sufficiently gross to cause serious confusion in a world where signal lights, colour television, photographic colour reproduction, and the matching of different batches of paints, dyes and textiles are specifically arranged to satisfy requirements of a person with normal colour vision. It is an example of the amazing adaptability of the human brain that these defects cause so few problems in everyday life. It is even true that adults are frequently unaware of colour anomalies which they have experienced from birth until they are confronted with the evidence of moderately sophisticated laboratory tests. Since an individual has no idea of the actual colours seen by any other person, apparent colour confusions are not readily discovered by verbal discussion. A colour defective who confuses red and green may recognize a traffic signal colour by its luminosity and position. Experience has taught him to call the darker colour red and the lighter colour green even though no true difference of hue may be seen.

10.2 A MODEL OF THE COLOUR VISION PROCESS AND THE POSSIBLE TYPES OF DEFECT

Figure 4.11 (p. 76) shows the possible normal organization of the initial stage of colour transmission. Many such models have been suggested and, as with the one chosen here, they assist in discussing and understanding the different types of colour defect which occur. Probably at least the initial stages of colour analysis occur in the retina but more central stages of the visual pathway may also be important. The receptors and circles represent neurons, which respond to wavelengths in the short, medium or long wavelengths (B, G and R respectively) or some combination of these. Beside each neuron is shown its spectral response (continuous line) and the three response curves of the neurons in the first stage (dotted lines) are included to assist comparison.

The trichromacy of normal colour vision probably results from the three types of cone pigments and their associated spectral responses are indicated by B, G and R in Fig. 4.11. The rod response is included but it will be assumed that this does not contribute to the perception of colour. Post receptor stages are shown by the luminosity neuron L (a summation of the $R+G+B$ responses), a Yellow neuron (a summation of $R+G$) and two colour difference neurons showing opponent responses to R and G, and Y and B, respectively. At the receptor stage colour information is transmitted by the B, G, and R neurons in a simple three-channel system. At the final stage of this model the same information is encoded in the output of the two-colour difference neurons and the luminosity neuron, hence this remains a three-channel system. The rod receptor is assumed to encode luminosity information only and to operate at retinal stimulations below those required for the three-colour channels, although there is evidence that it still operates at quite high luminances.

A number of possible types of defect can be predicted from this model:

1. Either the three-cone pigments are missing or they do not convey information to the later stages resulting in only rod vision (the rod monochromat).

2. Two of the types of cones malfunction resulting in single cone vision (the cone monochromat). It is possible, if not probable that all three B, G and R responses still contribute to L giving a normal luminosity response, but not to the colour difference channels.

3. One of the three-cone systems malfunctions resulting in the cone dichromat. The luminosity function may be normal or result from a summation of only two of the cone responses.

4. The R–G system malfunctions resulting in a normal B–Y colour difference channel and a normal luminosity channel. This type of dichromatic vision can be attributed to a fusion of the R and G responses and not to a loss of a single cone response.

5. Other defects caused by malfunctions in the B–Y neuron, the Y neuron, or L neuron, or combinations of these and other defects can be imagined. Most established colour vision defects fit into the predicted types 1 to 4 above, however.

10.3.1 Normal colour vision

The normal subject can match the appearance of a spot of light imaged on the fovea of any colour and spectral composition by a suitable mixture of three primary colours. As was shown in Chapter 7 (equation 7.1) this can be expressed by the trichromatic equation:

$$D(D) \equiv A(A) + B(B) + C(C),$$

where (D) represents the colour to be matched in appearance ($\equiv$) by A,,B and C units of the primaries (A) (B) and (C). The normal algebraic laws are obeyed. (The symbols A, B, C and D are synonymous with C, R, G and B of equation 7.1, p. 112.)

To a useful approximation all normal subjects require the same amounts of the primaries to match the chosen colour. However, some subjects who are classified as normal do often exhibit small but consistent differences. These are explained by differences in the spectral absorption properties of the lens and also by the yellow Macular pigment which covers the central 5° to 10° of the retina. The lens tends to become more yellow with age. There is evidence that the Macular pigment absorbs principally in the shorter wavelengths with a density which varies between subjects. Since both these effects modify the light before it reaches the retina the variations they cause are not usually included as representing defective vision.

It is unlikely that normal colour vision properties vary with race and other population characteristics although most of the available colour vision data have been obtained from subjects in North America and Europe. There is evidence of differences in the lens pigmentation for different races which may result in colour vision differences at a pre-retinal level. It is also interesting that the incidence of the more common colour vision defects has been claimed to be lower in non-Caucasian races. The incidence and types of defects based on European and American studies are shown in Table 10.1. Normal subjects cannot generally be induced to experience colour vision defects for example by placing filters between the eye and a coloured field. With peripheral vision there are indications that the ability to discriminate colour is reduced, however. Of particular interest is that of observation with small fields of less than 20′ subtense. Here a normal subject tends towards a dichromat with vision not unlike the tritanope, i.e. with a reduced response to the blue wavelengths.

A subject with normal colour vision is called a normal trichromat because in general he requires three primary colours to make a colorimetric match. Similarly monochromats and dichromats require only one or two suitably chosen primaries. Those subjects which still require three primaries but make colorimetric matches which are not accepted by the normal trichromat are called anomalous trichromats.

10.3.2 Monochromats
Rod monochromats apparently possess no functional cones and con-

TABLE 10.1

PERCENTAGE FREQUENCY OF OCCURRENCE OF COLOUR VISION DEFECTS

Type	Male	Female	All
1. Monochromats			
(*a*) Rod	0·003	0·002	0·0025
(*b*) Cone	small	small	small
2. Dichromats			
(*a*) Protanope	1·2	0·02	0·61
(*b*) Deuteranope	1·5	0·01	0·76
(*c*) Tritanope	0·002	0·001	0·0015
3. Trichromats			
(*a*) Protanomalous	0·9	0·02	0·46
(*b*) Deuteranomalous	4·5	0·38	2·5
(*c*) Tritanomalous	small	small	small
Total	8·1	0·43	4·3

sequently see no colour but simply achromatic variations of light and shade. Their vision is similar to that of the normal subject under low retinal illuminances, that is they exhibit poor acuity and have a normal scotopic relative luminous efficiency response. They are also photophobic, showing signs of discomfort to medium and high levels of light.

Three types of cone monochromat might be expected due to a loss of any two of the three cones. Those cases which have been studied seem to possess either green or blue cones only. These cases are rare, however (see Table 10.1) and it is not certain that this is a receptor or more central malfunction. The rod function is generally normal.

10.3.3 Dichromats

Three types of dichromat have been established which is as expected if one of the cone channels is not present or functions abnormally. The evidence suggests that the protanope (pro- meaning first) lacks the red sensitive cone pigment, (R), the deuteranope (deutero- meaning second) the green (G), and the tritanope (tri- meaning third) the blue sensitive pigment (B). It is also possible (see Fig. 4.11) for the red and green cone responses to fuse to form a yellow (Y) channel but for no output to arise from the R-G colour difference channel. This could be due to the red and green pigments being combined in single cones or due to a more central fusion. Such a model has been used to explain deuteranopic vision.

10.3.4 Anomalous trichromats

An interesting feature of the dichromat is that he will accept colour matches made by a normal subject although as described below he will confuse many colour matches which the normal readily distinguishes. This suggests a simple loss or fusion at some stage in the colour discrimination channels. The anomalous trichromat, however, while requiring three primaries to complete a colour match, uses quantities of the primaries which are significantly different from the normal. This defect cannot be explained by a simple decrease in sensitivity of one of the three colour vision channels. Whereas the dichromat appears to suffer a loss in the visual system the anomalous trichromat suffers a change. To some extent, however, the anomalous subject seems to possess visual characteristics which are half way between the dichromat and normal trichromat.

Consequently the names protanomalous, deteranomalous and tritanomalous are used to indicate a reduced sensitivity, principally in the red, green and blue regions of the visible spectrum. Strongly anomalous subjects tend to have colour characteristics similar to the corresponding dichromat and the words protan deutan and tritan are often used to include both the corresponding dichromat and anomalous subject.

10.4 CHARACTERISTICS OF DICHROMATS AND ANOMALOUS TRICHROMATS

The characteristics of colour vision defects are most readily described by reference to those psychophysical functions which are dependent on a subject's ability to discriminate or respond to lights of different colours.

10.4.1 Wavelength discrimination

The wavelength discrimination function can be measured by placing two fields side by side both initially being illuminated by monochromatic lights of the same wavelength. If the luminosities are equal they will appear identical to all subjects. The wavelength of one field is now slowly varied until the subject reports that a difference in appearance is just noticeable. (It is important to maintain equal luminosities during the change so that the threshold difference is caused only by the difference in wavelength.) A graphical plot of the reciprocal of the difference in wavelength ($\Delta\lambda$) at each wavelength (λ) is a measure of the discrimination sensitivity. For normal colour vision this sensitivity is low at the extremes of the visual spectrum where mainly one of the three cones is responding and maximal at wavelengths close to 490 nm and 590 nm where at least two of the cones will be strongly stimulated. (see Fig. 6.5, p. 104)

Figure 10.1 shows the wavelength sensitivity functions of dichromats for the three types protan, deutan and tritan. The dichromat shows no colour discrimination in the region of the spectrum where a single cone response function predominates. Hence, the protanope and deuteranope have no ability to distinguish monochromatic colours above approximately 550 nm and the tritanope behaves similarly for the shorter wavelengths. The similarity of the protanope and deuteranope functions can be explained by assuming that the red and green responses extend over a similar range of wavelengths but other factors may be involved. It would be of interest to know which colour each dichromat sees at the region of no discrimination but as mentioned previously, reported descriptions can be misleading. A few subjects have been found who possess dichromatic vision in one eye and normal vision in the other. These reports are open to question but the protanope probably sees all long wavelengths as green and the deuteranope all long wavelengths as yellow. This is as expected if the protanope suffers a loss of the red system and the deuteranope a fusion of the red and green systems.

Anomalous trichromats exhibit functions somewhere between the normal and the dichromat. The reduction in discrimination can vary considerably between subjects and the wavelength sensitivity function can be used to detect the extent of the anomaly. These functions support the view that one of the three response channels is not only different from the normal but suffers a reduction in sensitivity.

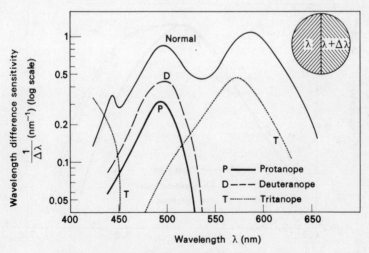

Fig. 10.1 Wavelength difference sensitivity function for subjects with normal trichromatic and with dichromatic vision.

10.4.2 Relative luminous efficiency function (V_λ)

The hypothesis that the protanope has a malfunction of the red channel is supported by the fact that more energy is required to see long wavelength stimuli. This is shown in Fig. 10.2 by a reduction in V_λ or the photopic relative luminous efficiency (section 3.5, p. 45) at long wavelengths and a shift of the maximum to the shorter wavelengths. The deuteranope function appears to be similar to that of the normal which can be explained in two ways. Either, the green channel is lost, but is similar in spectral response to that of V_λ so that the decrease in relative sensitivity is similar at all wavelengths. Alternatively, the red and green channels are fused to give a function similar to that of the normal. At the present time the evidence is equivocal and it may be that two types of deuteranope exist, one whose defect is due to loss and one which is due to a fusion. The tritanope function shows a small decrease in sensitivity at the shorter wavelengths but this may not be significant. This is perhaps simply an example that the blue channel contributes relatively little to the perception of luminosity.

Anomalous trichromats have V_λ functions similar to the corresponding dichromat although the corresponding decrease in sensitivity is sometimes smaller. This again indicates that in many respects the anomalous colour characteristics lie between the normal and the dichromat although the origins of the two types of defect are probably not identical.

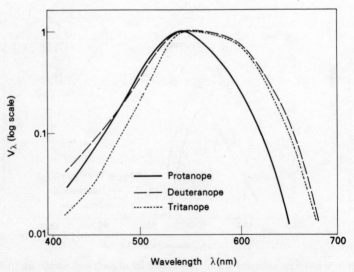

Fig. 10.2 Relative luminous efficiency function (V_λ) for dichromats.

10.4.3 Purity discrimination

To some extent the wavelength discrimination function and the V_λ function show the variation of hue discrimination and the luminosity response to different wavelengths. The ability to discriminate the degree of saturation of a field from white is similarly shown by the purity discrimination function (Fig. 10.3). This function is measured in a similar manner to the wavelength discrimination function. Now, however, the two fields are initially set at white and a monochromatic light slowly added to one until the change in purity is just noticeable. Again it is important to keep the luminosities equal at all times. The purity difference Δp is defined as the ratio of the added coloured luminance to the total luminance. (see also Fig. 6.6, p. 105)

The minimum at 570 nm in the normal function is perhaps indicative of the low saturation of spectral yellow when compared with other wavelengths. Protanopic and deuteranopic subjects similarly have minima at approximately 490 nm and 500 nm respectively. In this region a change from a white field by addition of one of these wavelengths does not change the colour appearance hence these monochromatic colours appear achromatic to the dichromat. A similar phenomenon occurs with anomalous subjects to an extent which depends on the degree of anomaly.

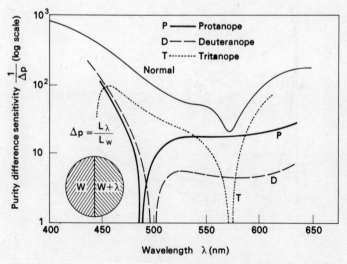

Fig. 10.3 Purity difference sensitivity function for subjects with normal trichromatic and with dichromatic vision.

10.4.4 Colour mixture and the chromaticity diagram

A dichromat accepts the colour matches of the normal subject but confuses colours which are easily distinguished by the normal. These confusions include all those sets of colours which are normally only distinguishable by the effects they have on the malfunctioning colour channel. This is most readily shown on the chromaticity diagram of the normal subject. Figs. 10.4, 10.5, and 10.6 show the loci of the confusion colours for the protanope, deuteranope and the tritanope respectively, on the CIE chromaticity diagram. As expected these lie on straight lines. That line which passes through white also cuts the spectrum locus thereby defining the point in the spectrum which appears achromatic, i.e. identical with a white of the same luminosity. These so-called neutral points occur at approximately 490 nm, 500 nm and 570 nm for the protanope, deuteranope and tritanope, respectively. The exact values will vary with the chosen white.

Anomalous trichromats do not generally accept colour matches by the normal subject. They have reduced colour discrimination as indicated

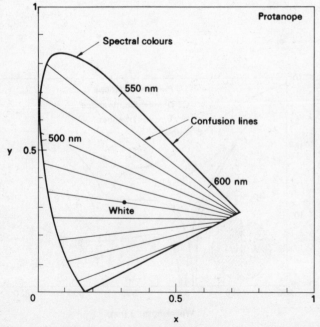

Fig. 10.4 Colour confusion loci for protanopes plotted on the CIE (xy) chromaticity diagram.

by the purity and wavelength discrimination functions. Generally these effects are not as large as for the dichromat and complete confusion lines do not exist. However, the MacAdam ellipses (section 6.5.1, p. 107) of the anomalous subject are distorted approximately in the direction of the corresponding dichromat's confusion lines.

10.5 DETECTION OF COLOUR VISION DEFECTS

The detection of colour vision defects is important where a person is performing a task which depends for its success on colour discrimination. Many tests have been developed some which simply screen or separate all colour defectives from the normal and some which attempt to diagnose the type and degree of the defect.

10.5.1 Nagel anomaloscope

The best known instrument for distinguishing protan and deutan defects is the anomaloscope first developed by Nagel in 1907. A monochromatic yellow field is matched in appearance by an additive mixture of mono-

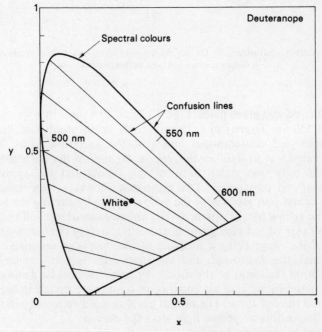

Fig. 10.5 Colour confusion loci for deuteranopes plotted on the CIE (*xy*) chromaticity diagram.

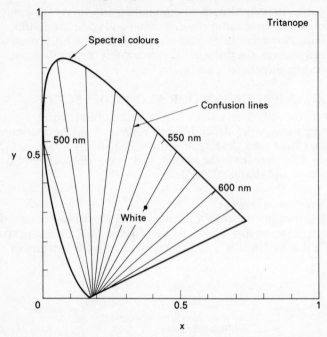

Fig. 10.6 Colour confusion loci for tritanopes plotted on the CIE (*xy*) chromaticity diagram (data not well authenticated).

chromatic red and green fields. Figs. 10.4 and 10.5 show that wavelengths between 550 nm (green) and 700 nm (red) lie on confusion lines for protanopes and deuteranopes and that the spectrum locus for the normal subject is straight in this region. Therefore the normal subject can make only one match between the yellow and the appropriate mixture of red plus green. The dichromat, however, can match any mixture of red plus green with the yellow simply by varying the luminosity of the yellow (or the mixture). The anomalous subject will be able to match a range of red plus green mixtures by varying the luminosity, the extent of the range being a measure of the degree of anomaly. Hence the normal, the dichromat, and the anomalous subject can be distinguished from the range of the match. Protans can also be distinguished from deutans by making an equality of luminosity match between the yellow and the red alone. The protan being insensitive to red will require a lower luminosity of yellow light than the deutan.

This test is the definitive test for protan and deutan defects but will fail to recognize the tritan who behaves similarly to the normal subject when matching red plus green mixtures. Since no region of the spectrum locus follows the tritan confusion lines a similar test cannot be derived using monochromatic fields. However, the difficulty can be overcome by using carefully chosen non-spectral stimuli as in Pickford's version of the anomaloscope.

10.5.2 Pseudo-isochromatic charts

As we have seen previously even a subject with normal colour vision cannot distinguish all mixtures of monochromatic wavelengths. The simplest example is the additive mixture of red and green lights which is seen as yellow. The colour defective distinguishes a fewer number of lights than the normal. It is possible therefore to produce coloured surfaces, reflecting a mixture of many wavelengths, which appear different in colour to the normal subject but appear identical to the colour defective. Diagrams containing such colours are often called pseudo-isochromatic charts.

These charts have arrangements of different coloured dots which appear to form some pattern or symbol. Colours are chosen which lie on the confusion loci of dichromats and anomalous trichromats. Consequently certain symbols (groups of dots) are seen as a single colour to the defective subject but not by the normal and vice versa (Plate 10, *bottom*). A number of such tests exist but few are suitable other than for a general screening of protan and deutan defects. For this purpose they have the advantage of cheapness, however, and of providing a rapid test which can be quite enjoyable for the subject.

10.5.3 Hundred hue test and the hue circle

The Farnsworth–Munsell hundred hue test is based on the poor colour discrimination of protans, deutans and tritans in specific regions of the Munsell hue circle (section 6.3.2, p. 96). The subject has to arrange in order of hue a number of chips* of equal Munsell chroma and Value in the form of a colour circle. Fig. 10.7 shows those regions of the colour circle where the defective subject frequently places the chips in a different order to the normal subject. The number of departures from the normal arrangement gives additional information on the severity of the defect. While this test is very sensitive it will obviously take several minutes to complete and perhaps as long to analyze. Consequently similar tests based on the same colour circle but involving the use of fewer chips are also used where rapid screening of large numbers of subjects is required.

An analogous test based on purity discrimination as opposed to wavelength discrimination depends on locating the neutral point (section

---

* The original test contained 100 chips but this was subsequently reduced to 85 to make the intervals more uniform.

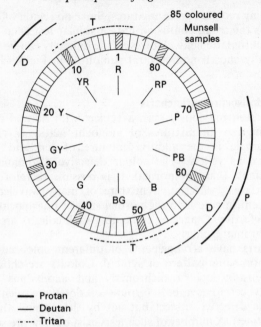

Fig. 10.7 The 85 colours of the 100-hue test showing the regions of abnormal errors for protans, deutans and tritans.

10.4.4, p. 162) in the hue circle. If the hue circle is observed at different purity levels by slowly adding white to each of the hues, anomalous trichromats and dichromats can be distinguished as well as the type of defect. A test based on this principle which has recently been developed by Dain is known as the Lovibond Colour Vision Analyzer.

10.5.4 Colour vision tests for specific tasks

Other colour vision tests are available which are related to specific tasks. For example the Lantern test is useful for testing subjects whose occupation involves recognizing signal colours, such as train drivers, pilots, etc. A series of glass filters is placed in front of a tungsten lamp and the subject is asked to describe the colour. The field size can be varied to simulate different viewing conditions. The filters and field sizes are chosen to relate realistically to the working conditions.

10.5.5 Objective tests

Each of the tests described above depends a great deal on the co-operation of the subject. Ideally it would be preferable to use an objective test which directly monitors the state of the visual system. This would be particularly useful with young children who cannot always co-operate in tests requiring concentration for more than a few minutes. At the present time such tests are only available in the research laboratory where extensive measurements can be recorded and these are not useful for the colour screening of large groups of subjects.

A colour defect arising from a loss of one of the cone pigments can be detected in theory by the technique of retinal densitometry. A beam of light is projected onto the fovea and the amount reflected measured with a sensitive photo-electric cell. The amount of light absorbed by the retinal pigment can be measured by comparing the amount of monochromatic light reflected by the unbleached or dark adapted eye with that reflected by the bleached or light adapted eye. If this is carried out at selected wavelengths above 500 nm the absorption curve due to the red and green receptors can be detected. If the two pigments are present a complex function is obtained but if one is present this function is relatively simple and its form independent of the colour of the bleaching light. Rushton has suggested, from measurements of this kind, that the protanope and deuteranope each lack a foveal pigment compared with the normal. However, at present only a few subjects have been tested and it is possible that two types of deuteranope based on a loss or fusion of colour-vision channels exist.

Other techniques which show promise for detecting colour vision defects are based on electrophysiological measurements using electrodes placed externally to the eye and brain. The Electroretinogram (ERG) (section 2.8, p. 32) where an electrode is placed on, or close to the cornea, monitors the retinal functions. More central visual processes can be monitored by the Electronencephalogram (EEG) (section 2.8, p. 33) with electrodes placed on the scalp near the so-called visual area. Both these techniques have been used to measure functions which contain information on hue discrimination and relative luminous efficiency. Hence colour vision defects may be evaluated as in the comparable psychophysical experiments. At present the ERG and EEG are mainly restricted to research applications but they have been extensively used for clinical evaluation of other visual characteristics and may in the future play a role in colour vision testing.

10.6 CONGENITAL COLOUR VISION DEFECTS

One of the interesting features of Table 10.1 is the large difference in incidence between male and female subjects; at least for the protan and deutan defects. This and the increased incidence of such defects in successive generations of some families point to the genetic basis of some colour defects. A human being possesses 23 pairs of chromosomes, one pair of which carries information relating to sex characteristics. The female has two similar sex chromosomes known as XX and the male two dissimilar sex chromosomes XY. The transmission of a colour vision defect depends on the gene carried in the X chromosome. A colour vision defect only becomes apparent when the full complement of X chromosomes are affected. Consequently a female with one normal X and one defective X does not exhibit defects herself but can act as a carrier. The male inherits the X chromosome from the mother and the female one each from the mother and father. Assuming that a defective X chromosome has no effect on survival of the species the proportion of colour defective females in the total population should approximate to the square of the proportion of defective males. This is seen to be approximately true if the total protan and deutan values are analyzed separately (Table 10.2).

10.7 ACQUIRED COLOUR VISION DEFECTS

Sometimes pathological changes due to external sources induce a colour vision defect in the originally normal subject. These may show effects similar to congenital defects, a general progression occurring from normal trichromatism through a form of anomalous trichromatism to dichromatism and finally perhaps monochromatism. Frequently these defects are obscured by other visual defects and this analogy must not be taken too far. Drugs used in the treatment of non-ocular diseases have been known to induce colour vision defects which are sometimes irreversible. The excessive consumption of tobacco, alcohol and stimulants have also been associated with a deterioration in colour vision. Ocular diseases such as glaucoma, ocular hypertension and in particular those diseases which affect macular or central vision have associated colour defects involving normally a deterioration of red-green or blue-yellow discrimination. A deterioration of colour discrimination associated with ageing of the lens has been mentioned previously in section 10.3.1, p. 156.

TABLE 10.2

COMPARISON OF PROTAN AND DEUTAN PERCENTAGES FOR FEMALES AND PRE-DICTED FEMALES

Type	Male	Predicted Female*	Female
Protan	2·1	0·044	0·04
Deutan	6·0	0·36	0·39

* Square of male proportion.

10.8 SOCIAL AND OCCUPATIONAL IMPLICATIONS OF COLOUR VISION DEFECTS

Colour defective persons lack an attribute of perception which may take on varying degrees of importance. The cone monochromat is obviously at a severe disadvantage in situations where colour is used to facilitate the communication of visual information. Dichromats and anomalous trichromats will generally suffer to a lesser extent. We are all familiar with black and white photographs, films and television, and our acceptance of these is indicative of how many aspects of visual appreciation are independent of colour. The amount of visual information which can be contained in line drawings and cartoons further indicates that in most circumstances the communication of form, luminosity and colour are in this order of decreasing degrees of importance. Colour is often a luxury rather than a necessity.

In the world of art and fashion, where colour is often deliberately used to stimulate, the colour defective must in general suffer a disadvantage. However, even here there will be examples where the defective subject sees the world in a more stimulating manner. Several world-renowned artists are now thought to be, or have been, colour defectives and it has been suggested that their anomalous view of the world has been used to advantage. A crude example of a defective subject using his anomaly to advantage is when a pattern is seen in a pseudo-isochromatic plate which goes unnoticed by the normal subject (Plate 10 (*bottom*)).

The occupational hazards of colour defects must obviously be avoided if possible. Consider for a moment the electrician who is unable to distinguish the red, green and black main supply cables which were of standard use in the United Kingdom for many years. A transport driver frequently has to recognize a coloured signal from too great a distance to rely on additional positional clues. In such occupations appropriate screening tests are a necessity to protect both the employee and others when safety is dependent on the efficiency with which the task is completed. Another field of employment which is restricted to the colour normal is that which involves the visual inspection of batches of goods of nominally the same colour. The colour normal is extremely sensitive to small colour differences and often finds slight variations in colour between different parts of an article which are obviously meant to be identical in colour most irritating. For example the door of a motor car may appear a different colour from the rest of the vehicle due to the use of more than one batch of paint of nominally identical colours with inadequate colour control. Similar examples can be quoted from the textile industry, etc.

Colour coding is frequently used to assist in the identification of apparently similar objects. Examples include the marking of pipes containing different fluids and gases in factories and the magnitude identification of electronic components such as resistors and capacitors. The colour defective may all too easily make important mistakes even when the defect is relatively mild. Where the person concerned is not aware of the defect he will often find fault with the type of colours used rather than blame himself. Consequently modern colour codes attempt to avoid the use of commonly confused colours or give additional non-colour clues where serious errors might be made.

In the field of colour reproduction such as in photography, the cinema and television it is important that the reproduced colours, which will at best be metameric matches (section 5.5, p. 83) of the original colours for the normal subject, do not appear unacceptable to the colour defective. In photographic reproduction this is of relatively small importance due to the use of broad band primary colours and the generally reduced colour discrimination of the colour defective. However, with the increasing popularity of colour television where narrow band colour phosphors are often used it may become important to consider the effects of colour anomalies, especially when the viewer becomes more experienced with, and hence possibly more critical of, the limitations of this form of colour reproduction.

11
Colour in our lives

11.1 WHY DO WE HAVE COLOUR VISION?

As we explained in Chapter 4, fish, birds, and the higher primates share with humans the ability to perceive images of their external surroundings in colour. Possessing colour vision means essentially that the organism can receive and handle a greater amount of information concerning its environment. It can thus distinguish ripe from unripe fruit such as strawberries and tomatoes. This capability obviously has a certain amount of survival value.

It is interesting that animals such as bees are attracted to certain flowers by their colour, and this is an example of behavioural liaison mediated by colour between the animal and plant kingdoms; the bees have the honey and the flowers are fertilized. However, since a very large number of species do not enjoy the advantage of colour vision and yet survive, one wonders why this incredibly complicated mechanism has evolved.

Aesthetically, of course, colour vision is very important to us and one would almost say it is an essential part of our everyday cultured life as we know it: But it is difficult to conclude other than that the aesthetic attributes of colour vision are a bonus and not a primary reason for the possession of a colour sense. One cannot imagine a bird or a fish appreciating a Turner or a Constable painting, or even finding delight in the coloured beauty of a terrain or a seascape.

Colour as an aesthetic experience is on a par with the appreciation of music, literature, scripture and art generally, and these are the spin-offs from the possession of an over adequate enormous brain which contains a great deal of redundancy. They lead to complex civilizations and cultures, which are quite rightly highly prized and rewarding, and eminently worthwhile, but which nevertheless do not possess any elements of basic survival.

11.2 COLOUR IN OUR SURROUNDINGS

In spite of these preceding remarks we live nevertheless surrounded by colour-resplendent nature. We have green grass beneath blue skies, and

171

a world containing brightly coloured flowers and animals. Furthermore, although we cannot survive by living in the open air, we often decorate the interiors of our buildings in the most lavish way. Even the humblest cottage has colour washed or coloured papered walls. We reflect our moods by the colours we wear in our clothes, and high office is often marked by the distinction of colour, for example purple.

11.3 COLOUR IN THE ATMOSPHERE

Let us consider for a moment the origin and purpose of some of the wonderful colours in nature. Why is the sky blue? This is a purely physically-based phenomenon. The tiny air molecules in the atmosphere scatter light. Lord Rayleigh found that the scattering was proportional to the inverse fourth power of the wavelength. This means that the blue light of short wavelength is scattered over eleven times more than the red light of longer wavelength.

Meteorological colour phenomena such as rainbows and haloes are similarly the result of the laws of physics concerning the action of matter on light rays. Rainbows are formed because light is refracted and reflected in a raindrop as shown in Fig. 11.1. The angle between the deviated rays and the rays from the sun is about 42°. This results in a reflected bow of light (Fig. 11.2) which is the locus of all raindrops which lie on a cone of semi-angle 42°. The fact that the refractive index of water differs with wavelength means that all colours are deviated differently, and the bow is broadened into a spectrum with the apex at the eye. The centre of the arc is at the anti-solar point, that is the point in the sky diametrically opposite to the sun. Since the sun must be on or above the horizon for a bow to be formed at all, the centre of the arc of the bow is either just

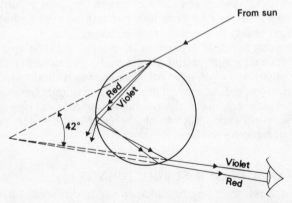

Fig. 11.1 The path of sunlight reflected in a water drop which forms a rainbow.

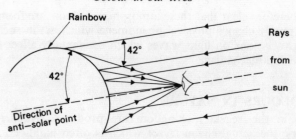

Fig. 11.2 The formation of a rainbow.

on or below the horizon. The reflection of the light within the drop results in a partial polarization in the rainbow. This can be proved easily by viewing a rainbow through a small piece of *Polaroid* which is then slowly rotated.

Sometimes a further secondary rainbow of larger radius (about 52°) can be seen outside the primary bow. This has the reverse sequence of colours to the primary bow, being red on the inside and violet on the outside. It is formed by light which has been reflected twice inside each raindrop, (Fig. 11.3). Similarly solar haloes and mock suns, and lunar haloes are formed when light from the sun or moon is refracted in ice crystals in the upper atmosphere. These are often only faintly coloured.

The most colourful natural phenomena after the rainbow are the Aurorae, which are due to charged particles ejected from the sun entering the Earth's atmosphere.

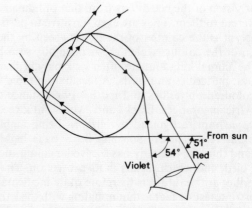

Fig. 11.3 The path of sunlight reflected in a water drop which forms a secondary rainbow.

It is therefore clear that these atmospheric colours are from our point of view purely chance physical phenomena which act in such a way to produce variations on light waves which incidentally affect the human colour vision mechanism.

11.4 COLOURS IN NATURE

Colours in the vegetable kingdom are produced by chemical light absorbers, the actual chemistry of which is often highly complex. The green colour of grass is due to chlorophyll, which has an absorption band in the red centred at 660 nm. This can easily be shown by extracting a solution of chlorophyll by mashing stinging nettle leaves with alcohol (or methylated spirits) and then filtering the liquid extract. This can then be placed in a transparent glass cell in front of the slit of a spectroscope. Using a source of white light the red absorption band can easily be seen.

The absorption of light by green plants is the basis of photosynthesis by means of which light energy is converted into chemical energy. This process is supremely important and it is that upon which all living organisms eventually depend. The fact that the grass looks green to us is incidental, and of little basic importance. Chlorophyll also reflects strongly in the infra-red and if our eyes were sensitive to these radiations, grass and leaves would appear much brighter and of a different colour.

Chemically all flower pigments contain a conjugated system of alternating double and single carbon—carbon bonds, usually with one or more chromophoric groupings. The most important are the anthocyanins, colouring from red to blue, the carotenoids, yellow, and the flavones, cream. Also both nitrogenous and quinonoid pigments occur.

Flower colours have a function in attracting honey-bees and other insects for pollination, which is essential to their life cycles. Since, however, the colour vision of the bee differs from that of humans, the colours will appear different to them. They are even sensitive to parts of the ultra-violet. Obviously it is the colours of the flowers seen by the bees which are important, not the colours which are incidentally seen by us.

Colours in the animal kingdom are often strikingly brilliant, especially in certain beetles, butterflies and birds. A fascinating topic is that of the marvellously coloured butterflies and moths. For instance the beautiful ostentatious African orange, black and white milkweed butterflies (*Danous chrysippus*) are recognized by birds as being highly poisonous since they contain certain heart poisons. Birds have highly developed colour vision, and thus the butterfly has evolved a colouration which can be seen by the bird, and which results in its protection. The fact that we think the butterflies are beautiful is therefore quite incidental.

A further very interesting fact is that mingling with the lethal butterflies are the so-called Batesian mimics which have evolved a striking resemblance to the *Danoids*. These are quite innocuous, but hiding behind their ingenious disguise, the birds leave them severely alone, since they mistake them for the poisonous species they imitate.

Animal colours are often not caused by pigments, for example, the colours of certain beetles such as *Heterorrhina elegans* are due to multi-layer thin film interference phenomena. Furthermore the blue colours of · the South American Morpho butterfly are due to three-dimensional diffraction in the complicated scale structure of the wings. The colours in the 'eye' of the Peacock's wing are produced similarly.

Nature thus uses very complicated chemical and physical means to produce colours which are more vital to the plants and animals than they are to us.

11.5 COLOUR IN OUR WAKING LIVES

Apart from a small convenience related to survival, such as being able to distinguish ripe from unripe fruit, it seems that the other advantages of our colour vision such as increasing the contrast and visibility of objects, are fairly marginal and not of vital importance to us.

Things are of course made more difficult at dusk and at night under dim conditions, since our retinal rod receptors give us neither good acuity nor colour vision. Nevertheless, it is possible to pursue quite a number of activities at night.

But what a contrast it makes to our sense of well-being to be in the broad sunlight of midday, rather than in the gloom of night. The eye can adapt to enormous changes of retinal illuminance, but it cannot see colours at night. Even in the day the true brilliance and saturation of colours is not realized until the sun comes out. Then we realise what colour means to our very being and to our spiritual and cultural existence.

It is interesting of course that we can often get by without colour. Most newspapers still only print pictures in black and white, and many films and television pictures are produced in monochrome.

Adding colour to illustrations and television screens, however, enormously increases the realism and enjoyment, and it is almost essential for a film for example on pictorial art such as painting, or for nature films, such as on flora and fauna.

It is difficult to escape from the conclusion that the aesthetic enjoyment of the colours of a seascape or landscape or a priceless work of art is the result of an almost accidental bonus to our senses of the possession of colour vision. It is on a par with the fact that our ears and brain are constructed in such a way that simply related physical vibrations in the air, when properly controlled, can give rise to the extremely pleasurable perceptions of melody and music. In other words, our eyes, our ears and our brain have evolved into such complex organization, that they are greatly over elaborate for the mere necessities of reproduction and survival. Thus their potential is immense. When man evolved beyond the point at which all his working hours were needed for survival, therefore, and when he had learned to extend artificially his naturally lighted hours, then this potential began to be tapped, developed and exploited. This was the beginning of the cultural spin-off of aesthetic and artistic appreciation.

Unlike all the rest of the animal and vegetable kingdom, our cultural existence sometimes predominates over our basic physical existence, and we rightly greatly prize our cultural and artistic heritage.

Colour therefore is one of the ingredients which has been instrumental in bringing about this great change in the perceptual dimensions of our lives. It immensely adds to the enjoyment and satisfaction of being alive. Perhaps this little study has helped to increase the understanding very slightly of this most fascinating and fundamental of enigmas.

11.6 COLOUR IN OUR SLEEPING LIVES

Our perceptions do not shut off entirely when we fall asleep. Modern research into sleep has revealed that there are two kinds of sleep the so-called normal, or orthodox sleep, and paradoxical sleep.

The latter is so named because during it, the individual exhibits brain and other bodily activities which are normally associated with our waking state.

It is during these paradoxical phases of sleep (perhaps three or more periods of about 25 minutes each night) that rapid eye-movements occur beneath the closed lids, and dreaming occurs. There is an in-built mechanism for forgetting most dreams, but when they can be re-membered they appear to be very real and lifelike experiences. However, surprisingly only about half of our dreams are in colour. Furthermore a number consist of black and white images with perhaps only one or two objects standing out in vivid colours. Why dreams should on the one hand be so realistic and yet often lacking in colour is obscure but may be due to colour being perceived by a different part of the brain to that perceiving luminance and form. The colour areas could sometimes be 'switched off' when dreaming.

With practice one can recall more of the contents of one's dreams, and it is often possible to remember the colours seen in dreams for some little time. One of the authors and some of his students recorded their dreams over a period of months, and as soon as possible after a remembered dream noted some of the colours seen with the aid of a Munsell colour atlas and other means.

The plots of their dream colours is shown on a CIE diagram in Fig.

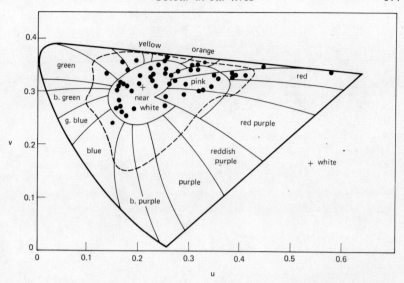

Fig. 11.4 Colours seen in dreams plotted on a CIE *u*, *v* chromaticity diagram. The dotted line represents the boundary locus within which lie the gamut of colours seen in everyday life which can be depicted by the use of pigments, dyes and inks.

11.4. Some interesting features emerge. On the whole, the colours seen are not highly saturated. There seems to be a predominance of reds and yellows, and a great dearth of blues. This is possibly due to the small numbers of blue perceiving units in the brain. Obviously exact colorimetry can never be done on the colours seen in dreams. It is an interesting point, however, that the colour perceptions in dreams cannot be basically different from those in our normal waking periods. The latter arise in the visual cortex from stimuli fed from the eye, whereas dream colours must result from the autostimulation of the same areas of the brain by activity of the brain itself.

To realize that colours can be seen after 'lights-out' is a good point at which to end our story, since it does not appear that we can proceed much further than this with the present state of our knowledge.

Research will go on and more and more pieces in the jig-saw of colour vision will be sorted out and put in their correct places. Perception and consciousness, however, are obviously centred on the brain, and it perhaps becomes a philosophical point to debate for how long and how deeply one can pursue a study of the inner workings of the brain using the brain to study itself.

Bibliography

Gregory, R. L. (1966) 'Eye and Brain, the psychology of seeing'. World University Library, Weidenfeld & Nicolson.

Helmholtz, H. von (1885–1895) 'Helmholtz's Treatise on Physiological Optics', vols I and II and vol III, Ed. J. P. C. S. Southall, Dover reprint, 1962.

Niven, W. D. (ed.) (1965) 'The Scientific Papers of James Clerk Maxwell'. Dover reprint.

Pirenne, M. H. (1967) 'Vision and the Eye'. Science Paperbacks, Chapman & Hall.

Weale, R. A. (1968) 'From Sight to Light'. Contemporary Science Paperbacks No. 24, Oliver & Boyd.

'Scientific American' published monthly often has excellent articles on vision. Single reprints of past articles can be obtained from W. H. Freeman & Co Ltd, 58 Kings Road, Reading, RG1 3AA.

For more advanced reading:

Cornsweet, T. N. (1970) 'Visual Perception'. Academic Press.

Hunt, R. W. G. (1974) 'The Reproduction of Colour', 3rd Ed. Fountain Press.

Judd, D. B. and Wyszecki, G. (1963) 'Color in Business Science and Industry'. Wiley.

Wright, W. D. (1969) 'The Measurement of Colour', 4th Ed. Adam Hilger.

Glossary of terms used in the text

Chroma (Used in the Munsell colour notation.) The chroma is the quality of a colour sensation which refers to the amount of pure colour present. It is related to the saturation (q.v.). A surface colour of high saturation has a chroma of 16 or more, whilst white has zero chroma.

Chromaticity The chromaticity is the colour quality of a stimulus expressed in terms of two chromaticity co-ordinates x and y on a chromaticity diagram. The x-value increases with increasing red content, and the y-value with increasing green content. There are three co-ordinates (only two being independent) which sum to unity, namely $x + y + z = 1$. The z-value increases with increasing blue content.

Colour Temperature When a perfectly absorbing body (i.e. a black body) is heated it emits light of a colour characteristic of its temperature, passing through reddish, yellowish, and bluish whites as the temperature is raised. Hence a method of specifying near-white colours is to quote the temperature in °K of the black body which has the same colour or a very similar colour.

Contrast (Subjective) The contrast is the subjective impression of difference of luminance or colour or both, between two adjacent areas.

Dioptre The power of a lens is measured by the reciprocal of its focal length in metres, and the unit is the dioptre. Thus a lens of focal length 1 m has a power of 1 dioptre, and a lens of focal length 0·5 m has a power of 2 dioptres.

Dominant wavelength The dominant wavelength with purity (q.v.) define the position of a colour stimulus on a chromaticity diagram in a similar manner to the chromaticity co-ordinates. A colour can be matched by mixing the appropriate amount of monochromatic light of the dominant wavelength with white light. Purples cannot be matched by adding monochromatic and white light together. With purples, therefore, the dominant wavelength is replaced by the complementary wavelength. (see Fig. 7.9, p. 121).

Hue Hue is the colour quality of a visual sensation which is expressed by different colour names such as red, yellow, green, blue, purple, etc. Physically it is related to the dominant wavelength.

Illuminance (or illumination) The illuminance of a surface is the luminous flux (or quantity of light per second) falling on unit area. *Unit*: lumen per square metre (lm m^{-2}) termed lux.

Luminance The luminance of a surface emitting or reflecting light is the luminous intensity per unit apparent area. *Unit*: candela per square metre (cd m^{-2}). It is a measure of the quantity of light per second emitted from a source or surface which reaches the observer.

Luminosity The luminosity is the quality of a surface or source which expresses whether it appears bright or dim.

Luminous flux Luminous flux is the rate of flow of radiant electromagnetic energy (or quantity of energy per second) evaluated according to its ability to produce visual sensation of light in the average human eye. *Unit*: the lumen (lm).

Luminous intensity The luminous intensity of a source of light is measured in candelas (cd). This expresses the quantity of luminous flux emitted in a cone containing one unit of solid angle (i.e. one steradian). Since there are 4π steradians in a complete sphere, a source of constant intensity 1 candela in all directions emits a luminous flux of 4π lumens. Luminous intensity is a measure of the illuminating power of a source, and of the amount of light it radiates to an observer.

Optical Density The optical density of a medium is defined as $\log_{10}(1/\tau)$ where τ is the transmission factor, or the ratio of transmitted to incident light. Thus an optical density of 1·0 corresponds to a transmission factor of 0·1, and an optical density of 2·0 to a transmission factor of 0·01. If two or more filters are placed in succession in the same beam of light the optical densities can be added.

Perception Perception is the actual conscious experience of the environment. It is based on the interpretation by the brain of incoming signals from the sense organs, in the light of previous experience which is stored in the memory. Visual perceptions enable us to experience the existence, colour, form and position of objects.

Purity The *excitation* purity represents the distance of a colour point from the white point on a chromaticity diagram compared with the distance of a spectral colour of the same dominant wavelength. Purity together with dominant wavelength (q.v.) define the position of a colour stimulus on a chromaticity diagram (see Fig. 7.9, p. 121). Colours, except purples, can be matched by a mixture of a monochromatic light and an appropriate amount of white light. The ratio of the monochromatic luminance to the white luminance is called the *colorimetric* purity. It is mathematically related to the excitation purity. For the same conditions of observation a high or low purity produces a high or low saturation.

Retinal illuminance The retinal illuminance is the luminous flux (or quantity of light per second) falling on unit area of the retina. *Unit*: lux, or more usually troland (td), where the number of trolands is the product of the pupil area of the observer's eye (in mm^2) and the luminance (cd m^{-2}) of the surface viewed.

Saturation The saturation of a colour sensation is the impression of the amount of pure colour present. The maximum saturation is perceived when a spectral colour is viewed at high luminance, and the minimum is when a white surface is viewed. It is expressed by the Munsell chroma (q.v.) for a coloured surface.

Sensation Sensation is the fundamental element of a perception. Thus we have sensations of light, colour, sound, temperature, pain, smell, taste, etc. Perception (q.v.) results when the sensation is interpreted by the brain.

Stimulus The stimulus is the factor in the environment (for example a source of light or of sound) which passes on energy to the sense organs giving rise to sensation and perception.

Value (used in the Munsell notation) The Value of a surface colour expresses the amount of light it appears to reflect. A true black surface has a Value of zero, and a true white a Value of 10. In this book it is used synonymously with the luminosity (q.v) of a surface.

Appendix

Information is given below on filters and colour samples mentioned in the text and on the suppliers from whom they can be obtained in the United Kingdom.

'CINEMOID' COLOUR FILTERS

Inexpensive, durable, non-inflammable 'Cinemoid' colour filters in the form of acetate sheets can be obtained from Rank Strand Electric, P.O. Box 70, Great West Road, Brentford, Middlesex, TW8 9HR. The reference numbers are:

Primary Red No. 6
Primary Green No. 39   } additive primaries
Deep Blue (primary) No. 20

Yellow No. 1
Magenta No. 13   } subtractive primaries
Turquoise (cyan) No. 62

A useful neutral filter is Pale Grey No. 60 which has an optical density of approximately 0·5.

HEAT ABSORBING FILTER

For experiments using intense light sources in which it is necessary to protect the eye from heat damage, a Chance–Pilkington Heat Absorbing Filter Glass HA 3, 4 mm thick can be used. This glass has a high absorption in the infra-red but a very high transmission in the visible region of the spectrum. These filters can be obtained from Precision Optical Instruments (Fulham) Ltd, 158 Fulham Palace Road, London W.6.

MUNSELL COLOUR SAMPLES

Munsell papers and Student Colour Chart Sets can be obtained from the agents, Tintometer Sales Ltd, The Colour Laboratory, Waterloo Road, Salisbury, Wiltshire.

For Maxwell's experiment with spinning discs the following colour samples can be used:

Red primary   Munsell 5R 4/14
Green primary   Munsell 5G 4/6
Blue primary   Munsell 5B 4/8.

Index

Plate 1

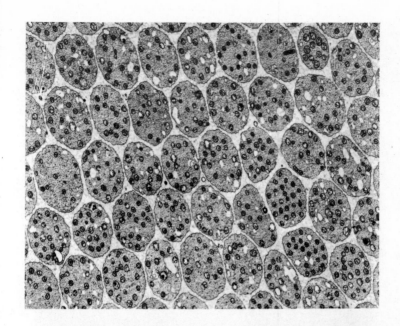

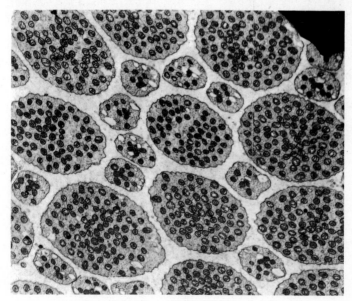

Retinal mosaic showing inner segments of the receptors. Monkey retina. (Prepared by G. Ruskell.)
(*top*) Cones at centre of fovea. Cone diameter 1·25 μm.
(*bottom*) Rods (smaller cells) and Cones near the edge of the rod-free area. Rod diameter 0·9 μm. Cone diameter 2·5 μm.
The dimensions of the receptors cannot be compared with those of human although these are similar.

Plate 2

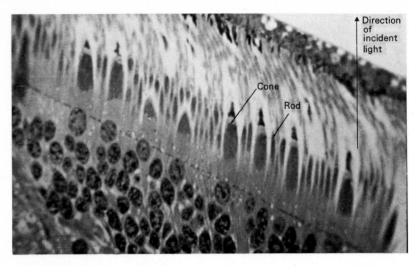

Transverse section of the retina near the optic disc showing rods and cones. (Prepared by G. Ruskell.)

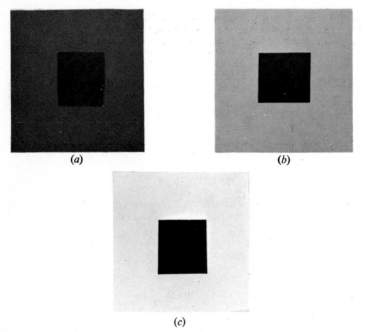

(a)   *(b)*

(c)

Changes in subjective contrast with change of background luminance *(a)* Low contrast; *(b)* Medium contrast; *(c)* High contrast.

Plate 3

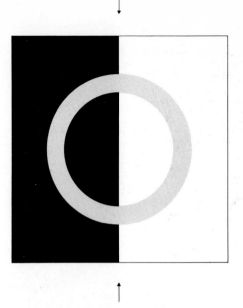

The effect of luminance contrast on luminosity. The grey circle appears brighter (less grey) against the black surround. This effect is greatly enhanced if a dividing line (a piece of black cotton or a pencil is laid across the arrow heads). Note the effect of moving this line slowly to the right or left.

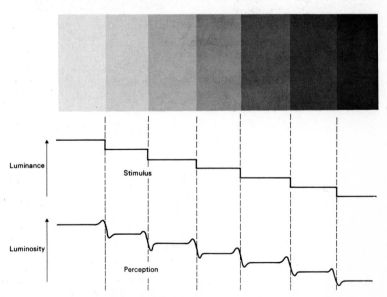

Mach bands. Each grey has a constant reflectance (indicated by the stepped stimulus) but the luminosity appears higher and lower either side of each change of grey (indicated by the curved luminosity function). This results in an enhanced contrast at each dividing line.

Plate 4

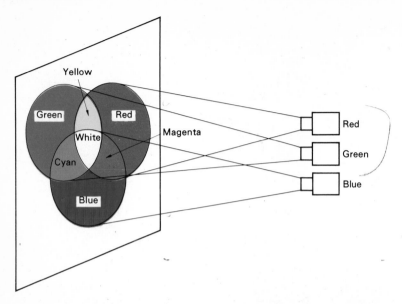

Thomas Young's colour mixing experiment.

Plate 5

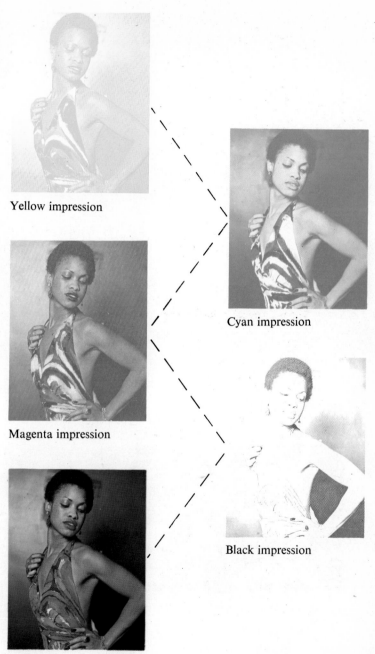

Yellow impression

Cyan impression

Magenta impression

Black impression

Final colour reproduction

Three colour separation positives and a black impression shown separately and superimposed to produce the final colour reproduction. (Photograph by Roger Slater.)

Plate 6

A half-tone photograph highly magnified.

A small area of a letterpress colour photograph highly magnified.

Plate 7

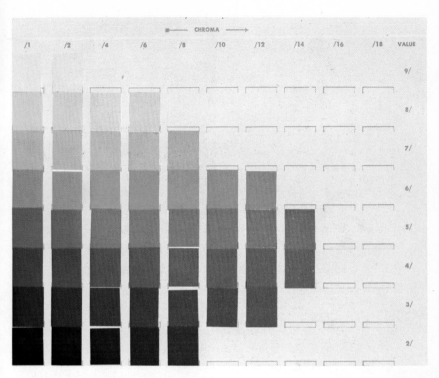

A page of the Munsell Glossy Book of Color for surfaces of the same hue (5R).

Demonstration of small field tritanopia.

Plate 8

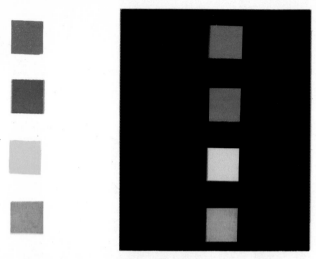

The effect of light and dark surrounds on the appearance of colours.

The effect of different colour surrounds on the appearance of colours.

Plate 9

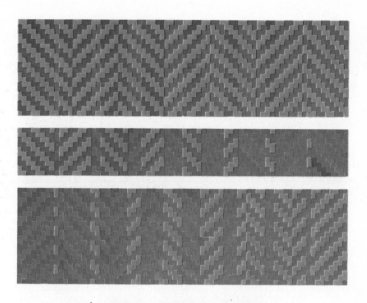

Simultaneous colour contrast. Design by S. Harry.

The ocular fusion of red and blue.

Plate 10

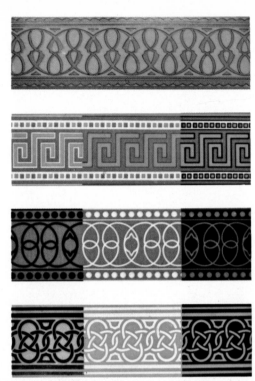

The spreading effect. Although the background colour is the same down each strip, when it is overlaid with a black pattern it appears darker; and when overlaid with a white pattern it appears lighter (from *An Introduction to Colour* by Ralph M. Evans).

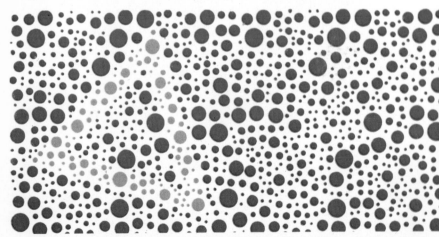

A pseudo-isochromatic plate. (Prepared by J. Birch.) A subject with normal colour vision will distinguish an orange triangle from the green background. A protanope or a protanomalous trichromat will not be able to distinguish the triangle or will be uncertain of its orientation.